Chemometrics: Chemical and Sensory Data

David R. Burgard, Ph.D.
Research Analytical Chemist
and
James T. Kuznicki, Ph.D.
Research Psychologist
The Procter and Gamble Company
Cincinnati, Ohio

CRC Press
Taylor & Francis Group
Boca Raton London New York

CRC Press is an imprint of the
Taylor & Francis Group, an **informa** business

First published 1990 by CRC Press
Taylor & Francis Group
6000 Broken Sound Parkway NW, Suite 300
Boca Raton, FL 33487-2742

Reissued 2018 by CRC Press

© 1990 by CRC Press, Inc.
CRC Press is an imprint of Taylor & Francis Group, an Informa business

No claim to original U.S. Government works

Library of Congress Cataloging-in-Publication Data

Burgard, David R.
    Chemometrics : chemical and sensory data / David R. Burgard, James T. Kuznicki
        p.   cm.
    Includes bibliographical references and index.
    ISBN 0-8493-4864-1
    1. Chemistry, Analytic—Statistical methods. 2. Sensory evaluation—Statistical methods.
  3. Chemistry, Analytic—Measurement. 4. Sensory evaluation—Measurement. I. Kuznicki,
James T. II. Title.
QD75.4.S8B87   1990
543'.0072—dc20   90-1893

A Library of Congress record exists under LC control number: 90001893

Publisher's Note
The publisher has gone to great lengths to ensure the quality of this reprint but points out that some imperfections in the original copies may be apparent.

Disclaimer
The publisher has made every effort to trace copyright holders and welcomes correspondence from those they have been unable to contact.

ISBN 13: 978-1-315-89150-7 (hbk)
ISBN 13: 978-1-351-07060-7 (ebk)

Visit the Taylor & Francis Web site at http://www.taylorandfrancis.com and the
CRC Press Web site at http://www.crcpress.com

PREFACE

The purpose of this book is to provide a practical reference manual for the interpretation of chemical and sensory data. Traditionally, unrelated scientific disciplines have developed data collection methodologies and procedures quite independently. Often the only common aspects have been the standard mathematical and statistical techniques used to analyze the data. This is the case for the disciplines discussed in this book: sensory science and analytical chemistry. Even though analytical chemistry techniques such as gas chromatography have been available for 2 decades, it was not until the advent of the capillary GC columns that enough detailed information, representing hundreds of different compounds, could be obtained on food and beverage items. Conversely, within the past 20 years the sophistication of sensory data collection techniques has increased their reliability to the point where they are also routinely a part of product development. The two areas have matured simultaneously and are increasingly used in combination. We therefore saw a need to provide an overview of both areas in one source book.

Both sensory evaluation and analytical chemistry have made use of various statistical tools. However, with the maturation of the disciplines and the increased volume of data capable of being obtained, increasingly sophisticated tools have been in demand. This demand is being met by the application of multivariate techniques. These methods can simultaneously compare and define the interactions of many variables and thus represent a way to mathematically integrate sensory and chemical data. Consequently, this book provides discussion of both sensory and chemical data collection, as well as examples of its use and combination in multivariate analyses.

Currently, there are numerous but separate sources which cover these topics in varying degrees of detail. These sources are referred to repeatedly and it is assumed that the reader can gain access to them. The focus of this volume is on the properties of the data itself, how they are influenced by methods of collection, and the effects those influences will have on interpretation. There is no attempt to go into great mathematical detail, although the mathematics are worked in places to provide an appreciation of the mechanics involved. The main thrust is toward interpretation rather than computation. We attempt to identify and discuss, but not resolve, some of the major issues that confront work in both of these areas. Readers interested in the development of mathematical derivations or more detailed discussion of theory will find ample references within the text. Also, we do not presume to be exhaustive in our coverage of the area. Rather, we intend to present representative examples of methodologies which provide the reader with an overall understanding of data collection methods as well as an appreciation for the multivariate approach. This volume should be particularly useful to a newcomer to the area of chemometrics, mainly because the major issues are identified and their impact on results is discussed. To this end we have attempted to keep the discourse as clear and direct as possible. The reader who has in-depth knowledge in one or both of the areas covered will probably notice where accuracy has been sacrificed for the sake of clarity to avoid being bogged down in difficult-to-understand details. The product development manager who must manage sensory or chemical data collection, or both, will also find this a source of material to help identify profitable avenues to pursue and problems to avoid.

This book is intended as a practical guide. We therefore emphasize what has been found in our experience to work, rather than what ought to work from the standpoint of any particular theory. However, it is also recognized that practice and theory are very often closely intertwined, and that theoretical ignorance can result in practical error. There is thus a certain compulsion to provide sufficient theoretical backdrop, especially where practice has been evolving independently of a theoretical framework to which it is clearly related. Importantly, some discussion of theory may stimulate the development of new application.

David R. Burgard
James T. Kuznicki

ACKNOWLEDGMENTS

We wish to thank the many individuals who provided support, encouragement, and understanding during the course of this work. In particular, we acknowledge the support of our families, Yvonne, Jessica, and Rachel, and Sue, Jason, and Matthew. Also, several individuals provided valuable technical support without which this work would not be possible. We wish to thank Ed Burton, Henry Dean, Brenda Keller, Al McDowell, Andrea Ngo, Don Patton, Lana Turner, and Lynn Work for collection of chemical and sensory data. Thanks are also due to Marcia Schulte for expert assistance with data analysis and to Eileen Fletcher for preparation of art work. In addition, we thank our colleagues at the Procter & Gamble Company for their encouragement and provision of the supportive atmosphere allowing this effort. Finally, we wish to acknowledge our indebtedness to the many colleagues and associates who have shared their knowledge and expertise with us over the years. Although too numerous to mention, it is our colleagues, teachers, and mentors from the chemical, sensory, psychophysics, and flavor communities who have truly provided the basis for this work.

THE AUTHORS

David R. Burgard, Ph.D., is an analytical research chemist in the Food and Beverage Technology Division of the Procter & Gamble Company. He received his doctorate in analytical chemistry from Purdue University in 1977. Since then, he has worked in pharmaceutical, food, and beverage research areas at Procter & Gamble.

Dr. Burgard's research interests include the development and application of analytical measurement techniques and the application of chemometrics. He has pioneered the application of chemometrics within the Procter & Gamble research organization. This effort has included the education of both management and fellow scientists as well as the direct application to product development research. He has published several papers and given numerous invited technical presentations on this research.

James T. Kuznicki, Ph.D., is currently a staff scientist in the Food and Beverage Technology Division of the Procter & Gamble Company. He received a doctorate in experimental psychology from the State University of New York at Albany in 1976 and did post-doctoral work in the Food Science Laboratory at the U.S. Army Research and Development Command at Natick, Massachusetts. He has worked in the food and beverage area at the Procter & Gamble Company since 1977.

Dr. Kuznicki's interests include psychophysics of taste and smell, the practical application of psychophysical techniques in a sensory evaluation context, the behavioral consequences and determinants of eating and drinking, and the development of preferences and aversions for both foods and beverages. He has published several papers concerning taste psychophysics and the perceptual processing of taste sensations. His work as a Sensory Evaluation Specialist and Project Manager have provided a broad range of experience in the application of techniques for collecting sensory and behavioral data as well as with various means of statistical analysis.

TABLE OF CONTENTS

Chapter 1

INTRODUCTION

I. GENERAL CONCEPTS

Most of the sensations that we experience in our everyday lives are complex. Quite often, we are not even aware of all the individual elements that combine to create the sensory impressions we experience. Current measurement methodologies in the sensory and physical sciences provide us with the tools to describe such complex phenomena in detail. In fact, we can now generate so much detailed information that it quickly overwhelms us. How then does one analyze the large volumes of data to find detailed information and identify relationships between chemical or physical measurements and sensory data? The application of systematic data reduction methodologies allows us to identify trends in large data sets and to study the interrelationships between data sets. Most of the techniques to be discussed are not new. The primary data analysis schemes include traditional univariate statistical and multivariate modeling algorithms.

The term chemometrics has traditionally been applied to chemical data. A similar application of many of the same techniques to sensory data is called psychometrics. For convenience, we will use the term chemometrics to include both types of data. When referring to sensory data, it is to be understood that psychometrics is implied.

A complete definition of chemometrics must include all aspects of data collection and data analysis. Massart, et al.[1] provide a good working definition: "the chemical discipline that uses mathematical, statistical and other methods employing formal logic (1) to design or select optimal measurement procedures and experiments and (2) to provide maximum relevant information by analyzing chemical data." As suggested by this definition, gathering the right data is as important as knowing how to analyze it. Therefore, we chose to cover the most important aspects of both chemical and sensory data collection and analysis. While the individual measurement techniques may vary, the general principles of data collection are applicable to all. Data collection aspects include the knowledge of experimental design, and instrumental and sensory measurements. Many of the same data analysis techniques are used in both the chemical and sensory disciplines, allowing the chemist to think about sensory information in familiar terms and vice versa. In fact, the chemist can consider the human as another type of instrument with which the sensory properties of chemicals are measured. No other instrument can provide such information.

For studies exploring the relationships between chemical and sensory data, measured variables are the amounts of individual chemical compounds in a sample and the sensory attributes that combinations of those compounds produce. In traditional terminology, the levels of the compounds would be considered independent variables and the sensory attributes the dependent variables. Whether dependent or independent, each set of measurements contain the following types of information:

1. Variable — variable correlations
2. Variable — sample relationships
3. Sample — sample comparisons

In all but the simplest system, the probability of correlations between measured variables is real. Identification of the interrelationships between variables and the lack of relationships is important to the understanding of synergistic as well as independent phenomena. Such knowledge helps the sensory scientist to evaluate the effectiveness of his panel to discriminate specific product attributes and to compare the performance of the individual panelists. With this information, he may be able to eliminate some measurement variables because they are redundant, or identify the need for additional variables in order to properly describe all the characteristics of a given product. In a similar manner, the chemist can study the information contained in chemical data. Organic chemistry tells us that, to a first approximation, compounds of the same chemical class (e.g., aldehydes) will undergo the same chemical reactions and should be affected in a similar manner by a given sequence of process steps. We would expect positive correlations to exist between related chemicals that are affected in the same way by processing. This information is often useful for functional classification of unknown compounds. Alternately, product and precursor relationships are suggested by the presence of negative correlations between two variables. A total lack of correlation between variables indicates the presence of independent phenomena.

Variable — Sample relationships help the scientist to begin to identify those variables indicative of certain characteristics of the samples being tested. This information also helps to verify that the samples in a test set possess the characteristics that were intended. We can all intuitively relate to the potential importance of the presence or absence of different compounds in a product. It is more difficult to understand and identify the significance when a gradual, simultaneous change in several variables is present across a series of samples. In fact, the simultaneous variation of several variables can create differences so complex that even the experienced flavorist may have difficulty understanding and describing them. Properly calibrated measurement tools, both chemical and sensory, can pick up trends, and analysis of the data can clearly demonstrate their presence.

Sample — Sample comparisons demonstrate similarities and differences in the characteristics of the samples tested. In combination with the previously mentioned relationships, the magnitudes of the differences needed to elicit a measurable change can be estimated. This ideally simplifies the task of controlling numerous variables simultaneously by allowing one to focus on those that are primarily responsible for the observed differences. For example, knowing that a twofold change in the levels of alcohols does not cause a measurable difference in the impression of a flavor system might allow the flavor chemist to concentrate on controlling the variation in the level of aldehydes which have a larger impact on the flavor characteristics.

When sensory and chemical data are combined, meaningful conclusions can be drawn. These might include the effects of processing or formulation changes on composition, the effects of composition on sensory characteristics and ultimately identification of the preferred composition. Obvious applications would be the prediction of the effects of variations in raw materials on flavor character, the identification and qualification of equivalent raw material sources, studying the effects of process changes on product performance, and understanding flavor stability. The development of information to support patent applications, to police patent infringements, and many other applications for predicting product attributes or performance from chemical or sensory measurements are possible. Many of the concepts developed in subsequent chapters can be applied in product development independent of the tools used to measure product composition and performance.

II. REQUIREMENTS FOR SUCCESS

A systematic approach to organizing and executing the various aspects of a complex study is absolutely necessary for successful implementation.[2] To begin, one must have a good appreciation for all the different aspects of the study. This includes knowing the specific questions to be answered as well as understanding the strengths and limitations of the techniques being used to gather, analyze, and interpret the data. Ideally, one person with the breadth and depth in all pertinent areas would be the most successful at solving such a problem. Realistically, a team of people, each with their own expertise and an understanding of the expertise of the others, is the best way to approach such work. This is simply because of the time consuming and tedious nature of the work required to generate meaningful information.

For successful implementation of such a project, the data must be reproducible and contain the information of interest. These criteria apply to both the chemical and sensory data. While they are obvious, they are not always met. Adequate attention to detail is needed in order for these requirements to be satisfied.

The reproducibility of the data (*precision*) is actually more important than the absolute *accuracy*, since the relative variations between samples are what is studied. This does not mean that accuracy in not desirable; rather, it means that similar conclusions can be drawn for both relative magnitudes as well as absolute magnitudes. This is primarily a result of the scaling techniques used for data preprocessing. We all need to remember that one can have good precision in the absence of absolute accuracy but cannot have accuracy without precision.[3]

Obviously, the data must contain the information of interest. Too often, the wrong data or only partial data are gathered. Attempts to analyze such data and draw meaningful conclusions are usually not successful and the poor results are interpreted to mean that the chemometric approach to data reduction is not capable of providing useful information. Detailed planning is essential to ensure that the right data are gathered. Analysis of data that have already been gathered for different purposes, or that were gathered without a specific question in mind, is often doomed to failure simply because the information needed to answer the questions of interest is not present. The fastest computers and the most sophisticated software cannot make up for poor data or the use of the wrong data. Trends cannot be found that are not inherent in the data.

Automation of data collection and handling is not an absolute necessity but certainly makes large studies easier to execute. The volume of data generated is typically large and difficult to manage manually. Human errors must be minimized.

III. THE CHEMOMETRIC APPROACH

The typical approach to a chemometric study is outlined in Figure 1. As indicated, studies are often not completed with one pass; rather, they are iterative in nature. The results of each iteration improve with the refinement of the understanding of the information in the data.

A. STUDY DEFINITION AND DATA COLLECTION

Generation of the database is the most important step. As mentioned earlier, the data must contain the information of interest. Therefore, the study needs to be planned in detail. Planning can begin by defining the study in terms of specific questions that are to be answered.[4] This approach helps one to determine what data need to be gathered, and if the questions can be answered simultaneously or must be addressed sequentially. Appropriate experimental designs must be considered.[4-6] Samples representing each characteristic of interest need to be included in the data set for comparisons. The samples should span the entire range of interest. It is dangerous to draw conclusions by extrapolation beyond the data set.

Figure 2 depicts some of the details that need to be considered in order to execute a complete chemometric study. While not all inclusive, the kinds of chemical, sensory and process information that might be considered during experiment planning are indicated. It provides a picture of the potential complexity of each part, the interactions leading to cause and effect relationships, and serves as a general guide to the types of information that can be useful. As indicated, it may be necessary to gather both chemical and sensory information with different techniques in order to adequately address all the characteristics in a given set of samples. Consequently, it may be appropriate to analyze subsets of the entire database with different data reduction techniques depending on the methodology used to gather the data and the experimental design.[5-7]

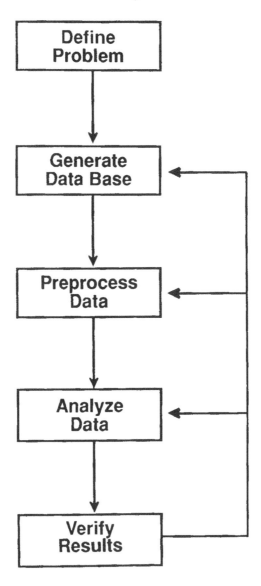

FIGURE 1. The chemometric approach. Several iterations are required to complete a study.

Initially, it is advisable to keep chemometric studies simple. This means that the researcher should limit the number of questions to be answered or the range of the sample characteristics to be studied. Smaller studies are easier to manage and the results easier to interpret. They allow one to develop familiarity with methodologies and confidence in the approach. As more experience and knowledge are gained, studies can be expanded by increasing the complexity of design or by combination of smaller data sets to answer several questions simultaneously.

More complete discussions about the size of the data base are contained in the discussions of specific techniques. However, it is instructive to mention some general concepts here. Most chemical and sensory data are not well behaved in a statistical sense. They may not have been collected according to a rigorous experimental design where all variables could be controlled appropriately. Such a situation arises as a result of numerous factors — complex variations in the chemical composition of natural products used as flavorings can be created by the growing conditions, variety of the root stock used, and processing. In addition, most flavor systems are

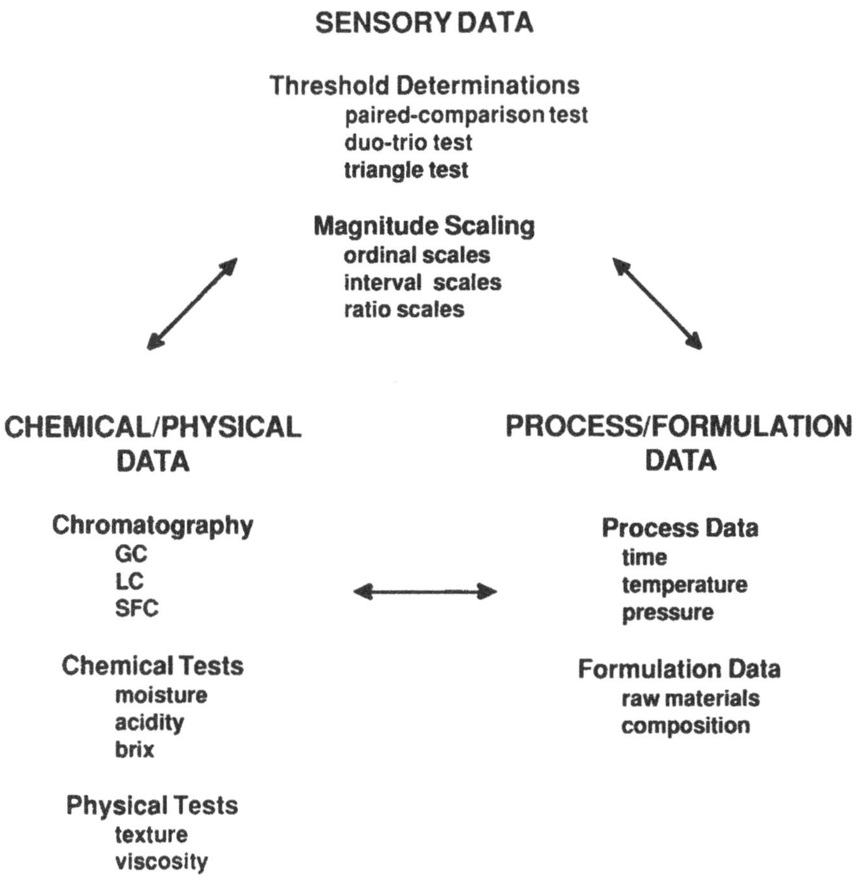

SENSORY DATA

Threshold Determinations
paired-comparison test
duo-trio test
triangle test

Magnitude Scaling
ordinal scales
interval scales
ratio scales

**CHEMICAL/PHYSICAL
DATA**

Chromatography
GC
LC
SFC

Chemical Tests
moisture
acidity
brix

Physical Tests
texture
viscosity

**PROCESS/FORMULATION
DATA**

Process Data
time
temperature
pressure

Formulation Data
raw materials
composition

FIGURE 2. Data collection details. Detailed planning for both samples and measurements is needed to ensure that all the necessary data is collected.

not derived from one source, adding more variables into the final product. These types of variations are inherent in the flavoring materials used for the formulation of the final product. Other characteristics of the data are introduced by the techniques used to analyze samples and gather the data. Modern chemical and sensory analysis techniques yield many pieces of detailed information about each sample. It is often difficult to gather data on enough samples to satisfy the criteria for application of traditional statistical procedures. Another potential problem exists because there are many *colinear,* or *correlated*, variables contained in the data collected for product characterization. Many of the correlations result from the biochemistry of natural flavors; others are caused by processing of flavor materials, chemical reactions during storage, and unknown or unexpected causes.

Multivariate techniques, like traditional statistical analyses, need enough data so that confidence in the results can be obtained. Unlike traditional statistical procedures, multivariate techniques are not typically designed to prove or disprove a specific hypothesis or to develop an unequivocal level of confidence. They are designed to identify trends and correlations in data and, as such, are not restricted by statistical sampling theories. Since multivariate analyses tend to be data condensing, they typically do not need as large a sample-to-variable ratio as univariate tests. In fact, the continuous modeling techniques, such as principal components, yield better results when there are a large number of related variables, as an averaging of the information contained in the numerous variables is obtained. This is analogous to the calculation of an average value from multiple determinations. From a practical point of view, the data-condensing and data-averaging characteristics of multivariate techniques make flavor studies feasible and allow one to begin with smaller data sets and to expand them as more knowledge is gained. Ultimately, any conclusions from such a study will have to be proven by gathering additional data, as discussed below in Section D on verification of results.

In the ideal situation, the exact data needed to answer a particular question can be gathered. However, the real world sometimes does not allow us to gather perfect data. We must learn to accept compromises. Accepting compromises does not mean that a good job is impossible; rather, it means that we need to understand the limitations imposed by the measurement techniques and be creative in finding ways to obtain the information of interest. The two most common limitations are the lack of samples that cover the entire range for a given characteristic and the inability of measurement methodology to provide adequate detail for complete sample characterization.

B. DATA PREPROCESSING

Preprocessing is probably the second most important step in a chemometric study. Preprocessing means to put the data into a meaningful form for further comparisons; that is, the conversion of raw data to units or scales that allow direct comparison of measurements for different samples. In the absence of preprocessing, erroneous conclusions can potentially be drawn. Preprocessing is often accomplished in three steps. The first step converts the data to units appropriate for the comparisons to be made, the second step organizes and creates a database, while the third mathematically conditions the data in preparation for the actual data analysis.

Sometimes steps one and three are done automatically or are transparent to the user and are not considered to be a separate step in an analysis sequence. For example, the conversion of chromatographic peak areas to parts per million can be accomplished in most instrumental data systems. This might be the first step in the preprocessing of chemical data. The third step, for example, centering and scaling, is often considered to be part of a standard data reduction sequence and is performed automatically by many software packages.

Step two is generally not available in commercial software packages and must be done manually or by user-created software or sequences. It involves aligning all the data for each sample to ensure that the same data points represent the same variable for each sample, and the entry of the data into the database in a form that can be accessed directly by the analysis software.

No matter how it is accomplished, preprocessing is needed to obtain the maximum amount of information from a data set and to prevent bias caused by improperly conditioned data. The user should determine exactly how data are being manipulated by any given analysis sequence. Detailed discussions of specific techniques are contained in subsequent chapters.

C. DATA ANALYSIS

Data analysis techniques are really systematic data organizers and reducers that allow one to search for and identify trends within a data set. Hopefully, the identified trends are meaningful, and are able to simplify data interpretation and provide insight into the answers of the original questions. Many different data reduction approaches are available. Even though the actual techniques can be very different, they all have several characteristics in common:

1. All are based on calculated measures of similarity between variables and samples.
2. All are data reducing in that they are used to identify the variables responsible for differences (and similarities) in samples; correlations between variables are usually identified as well.
3. All are predictive in nature and can be used to position unknowns with respect to known samples in an established database,
4. The user must interpret the results of the data analysis based on knowledge of the chemistry, sensory processes, and sample history.

5.   All identified trends are mathematically derived and *do not* necessarily guarantee cause and effect relationships.

D. VERIFICATION OF RESULTS

The last step in the sequence, the verification of results, begins with examination of the original data to ensure that the results of the data analyses have been correctly interpreted. Trends or relationships that are indicated by the data analysis should be observed in the data. This is best accomplished by visual inspection of graphical displays or tabulated data. Samples or variables that are *outliers* are usually found at this point. Many times, it is possible to trace abnormal behavior to errors in the database. Correction of the errors improves the quality of the results in the next iteration of analyses.

Any indications of cause and effect relationships need to be clearly demonstrated in subsequent experiments before conclusions can be finalized. Chance correlations must be considered a real possibility when working with large data sets. Chance correlations are not totally useless, as such information can potentially be used to select *indicator* variables for prediction of sample characteristics. For example, the appearance of hexanal and other aldehydes is often taken as an indication of lipid oxidation in food samples. While these aldehydes are not necessarily responsible for the changes in sensory characteristics, they are good predictors of the presence of offnotes in the samples. Such results may have some practical value, but do not answer the fundamental question of the causes of the offnote in the sample.

Cause and effect relationships are studied by gathering data for more samples and predicting the behavior of the new samples based on the data analysis models. Comparison of the predicted vs. measured behavior of the test samples adds confidence to the results, but still does not guarantee cause and effect relationships. To solidify cause and effect relationships, the modification of samples through flavor formulation may be required. Response surfaces of selected sensory response vs. formulated variations in the levels of identified compounds take the process one step closer to establishing cause and effect relationships. Two factor responses (two independent variables) are commonly used, as they allow the researcher to visualize the interactions of flavor chemicals. Mathematical models describing these relationships can be developed.

Validation experiments are not always straightforward, as it may be difficult to obtain the necessary samples. Difficulties include establishing the identity of trace level compounds (chromatographic peaks) responsible for the specific sensory character or even obtaining known compounds in adequate purity for spiking experiments. The ingenuity of the experimenter to identify and prepare the necessary samples is usually challenged at this point. Nonetheless, this step truly completes the study and is needed in order to clearly demonstrate cause and effect relationships between chemical and sensory data.

IV. BOOK ORGANIZATION

This book is organized into three sections. Chapters 2 through 4 discuss aspects of chemical measurement related to gathering good quality data. Similar insights into the fundamental characteristics of sensory data collection are covered in Chapters 5 through 7. The last three chapters are primarily focused on data analysis techniques and build on the information contained in the first seven chapters. A sample database, described in Chapter 8, is used to integrate the information in the three sections.

The chapters are designed to cover a given subject as a functional unit. The pertinent aspects, from theoretical background to practical data acquisition or reduction, are covered. This breadth of coverage is deemed important to provide adequate perspective on each subject. Due to the breadth, each topic may not be discussed in as much detail as needed for the inexperienced researcher to become proficient in a given area. The cited references should provide the detail necessary for one to gain such proficiency.

REFERENCES

1. **Massart, D. L., Vandeginst, B. G. M., Deming, S. N., Michotte, Y., and Kaufman, L.,** *Chemometrics: A Textbook,* Elsevier, Amsterdam, 1988, 5.
2. **Pardue, H. L.,** *Systems approach to food analysis, in Modern Methods of Food Analysis,* Stewart, K. and Whitaker, J., Eds., AVI Publishing, Westport, CT, 1984, chap. 1.
3. **Bevington, P. R.,** *Data Reduction and Error Analysis for the Physical Sciences,* McGraw-Hill, New York, 1969, 4.
4. **MacFie, H. J. H.,** Aspects of experimental design, in *Statistical Procedures in Food Research,* Piggott, J. R., Ed., Elsevier, Amsterdam, 1986, chap. 1.
5. **Massart, D. L., Vandeginst, B. G. M., Deming, S. N., Michotte, Y., and Kaufman, L.,** *Chemometrics: A Textbook,* Elsevier, Amsterdam, 1988, chap. 17.
6. **Gacula, M. C., Jr. and Singh, J.,** *Statistical Methods in Food and Consumer Research,* Academic Press, Orlando, FL, 1984.
7. **Meilgaard, M., Civille, G. V., and Carr, B. T.,** *Sensory Evaluation Techniques,* CRC Press, Boca Raton, FL, 1987.
8. **Wold, S., Abano, C., Dunn, W. J., III, Edlund, U., Esbensen, K., Geladi, P., Hellberg, S., Johansson, E., Lindberg, W., and Sjorstrom, M.,** Multivariate data analysis in chemistry, in *Chemometrics — Mathematics and Statistics in Chemistry,* Kowalski, B. R. Ed., D. Reidel, Dordrecht, Netherlands, 1984.

GENERAL REFERENCES

1. **Martens, H. and Russwurm, H., Eds.,** *Food Research and Data Analysis,* Applied Science, London, 1983.
2. **Teranishi, R., Flath, R.A., and Hiroshi, S., Eds.,** *Flavor Research — Recent Advances,* Marcel Dekker, New York, 1981.
3. **Schrier, P., Ed.,** *Flavour '81 — 3rd Weurman Symposium,* Walter de Gruyter, Berlin, 1981.
4. **Adda, J., Ed.,** *Progress in Flavour Research 1984,* Elsevier, Amsterdam, 1985.
5. **Martens, M., Dalen, G.A., and Russwurm, H., Jr., Eds.,** *Flavour Science and Technology,* John Wiley & Sons, Chichester, 1987.
6. **Charalambous, G. and Inglett, G. E., Eds.,** *Flavors of Foods and Beverages — Chemistry and Technology,* Academic Press, New York, 1978.

Chapter 2

SAMPLE PREPARATION AND EXTRACTION TECHNIQUES

I. INTRODUCTION

Chemometric studies require data for many samples. Generation of a good database takes considerable time and effort. Long-term planning is essential for successful implementation of a study. The particular sample preparation/extraction technique needed depends on the information desired, the sample matrix, the stability and levels of the analytes of interest, and the capability of the chosen instrumental technique(s). The preparation of samples for analysis can be more tedious and time consuming than the actual instrumental measurement.

Although good leads are available in the literature, preliminary studies of methodology characteristics may be needed to identify the most appropriate sample preparation technique. Practical considerations of time and sample throughput are also important. Selection of the sample preparation technique must include an evaluation of the merits of the individual techniques along with an understanding of any tradeoffs involved. Once an analysis procedure has been chosen and validated, the same scheme should be used throughout the study. It is risky to assume and often difficult to prove that an alternate analysis scheme will provide equivalent data.

II. SAMPLE PREPARATION TECHNIQUES

Summarized below are several of the more popular sample preparation/extraction methodologies. References 1 and 2 contain additional information on these and other sample preparation techniques. Table 1 compares the recoveries for several compound classes for the different techniques. It is impossible for such a table to cover all possible combinations of analytes, sample matrix, and extraction conditions. Therefore, these data are estimates of the range of recoveries possible. From the ranges reported, it can be seen that recoveries for individual compounds need to be determined for maximum accuracy.

A. SIMULTANEOUS DISTILLATION AND EXTRACTION (SDE)

This approach was first reported by Likens and Nickerson and has been widely used for the extraction of volatile flavor compounds.[3-5] SDE can be used for both liquid and solid samples. Solid samples are steam distilled from an aqueous suspension. Common extraction solvents are methylene chloride, a freon (e.g., freon 11), an alkane, or an alkane-ether mixture.

Figure 1 shows a diagram of a typical apparatus used for an SDE with a heavier than water solvent. Numerous variations of this apparatus are available. The volatiles are steam distilled from the sample, condensed with and extracted into the solvent. Flask A has the capacity for several hundred milliters of an aqueous sample or 1 to 10 g of a solid sample with 50 to 200 ml of water added. The sample flask is heated to above 100°C using an oil bath or heating mantle. Solvent (10 to 100 ml) is placed in flask B and brought to reflux. Flask B is heated with a water bath on a hot plate. The steam distillate and solvent are condensed on the cold finger, C, that is chilled with circulating ice water. Fresh solvent is continuously in contact with the sample distillate. The extraction occurs in the U-tube, D, where the solvent and water phase separate and are returned to their respective flasks through arms E and F. The Dewar flask, G, is filled with liquid nitrogen or dry ice to prevent the escape of any vapors. This scheme, when combined with solvent evaporation, can provide large concentration enhancements (up to 1000×). Extractions must run for more than an hour to exhaustively extract a sample.

The choice of solvent also depends on the analytes of interest. Theoretically, choice of the proper solvent can result in selective compound removal or the extraction of all compounds. Low boiling solvents make final extract concentration easier.

TABLE 1
Recoveries for Various Types of Compounds

Compounds	Range of recoveries (%)			
	SDE[a]	SDA[b]	SPE[c]	TH[d]
Aldehydes	60—100	34—71	44—93	70—80
Alcohols	>95	40—98	22—100	70—80
Esters	10—90	27—75	80—95	65—80
Ketones	60—100	45—100	30—100	70—80
Phenolics	55—100	nd[e]	85—100	nd
Pyrazines	>95	nd	50—98	nd
Terpeniods	>95	10—30	93—100	65—75

[a]   Data taken from several sources including References 3 and 4 and in-house experiments; ranges represent variations in operating conditions and samples.
[b]   Charcoal adsorbent; test mixture; Reference 6.
[c]   50 ml coffee, 2 g C_{18} packing, 2 ml $CH_2 Cl_2$ solvent.
[d]   25 ml orange juice, 5 ml $CH_2 Cl_2$ solvent, 30 s homogenization.
[e]   nd = not determined.

The rate of extraction follows a curve similar to that in Figure 2. A fixed time is chosen for the extraction. The rate of extraction varies for different compounds depending on their volatility, their solubility in the sample matrix, and the extraction solvent. Practical operating conditions will be determined by the rate of removal of the compounds with the lowest solvent-to-water partitioning.

The SDE technique is very good for qualitative analysis. It tends to be tedious and time consuming. Sample throughput can be improved by processing several samples in parallel. Extra effort is required to yield good quantitative data. Some concern has been expressed about artifact generation caused by holding the sample at elevated temperatures for extended periods of time. To minimize this problem, vacuum can be used to lower the boiling points and operating temperatures. Experienced operators can perform SDE extractions with relative standard deviations (RSD) as low as 5%. Typical RSDs fall in the range of 10 to 20% with 15% being a good practical target.

B. SIMULTANEOUS DISTILLATION AND ADSORPTION (SDA)

By putting an adsorbent trap in the condensate return path, the SDE technique can be modified to SDA.[6] In SDA, the analytes are steam distilled, condensed, and adsorbed on the trap (see Figure 3). The extract is introduced into the analysis instrument by desorption from the trap. Desorption can be thermal as described below for headspace analysis, microwave-induced,[6,7] or by solvent rinsing.[6] The characteristics of SDA are similar to SDE with respect to selection of operating conditions and sample throughput. Adsorption of analytes on the trap is essentially quantitative so the rate of extraction depends on the rate of distillation from the sample. Different adsorbents can be used for the trap, including various charcoal, Tenax, Porapak, or Chromosorb materials. The characteristics of several adsorbents are reviewed in References 1, 7, and 8.

C. SOLVENT EXTRACTION (LLE AND LSE)

Liquid-liquid extractions (LLEs) and liquid-solid extractions (LSEs) are the classical solvent extractions. Common devices range from the separatory funnel (for LLE) to the Soxhlet extractor (for LSE). Solvent extraction is both time consuming and tedious as multiple extractions are required to completely extract a sample. Devices such as the Soxhlet extractor must be allowed to operate for several hours to exhaustively extract the sample.

Ideally, solvents can be chosen to yield the selective extraction of certain compounds. In

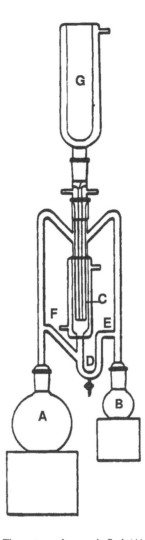

FIGURE 1. SDE apparatus. The parts are the sample flask (A), the solvent flask (B), the cold finger (C), extraction U-tube (D), solvent (extract) return arm (E), water return arm (F), and Dewar flask (G).

practice, extraction of all flavor compounds is desired and the opposite problem is encountered. The usable solvents do not provide a wide enough range of extraction for compounds of different chemical types. Commonly used solvents are pentane, hexane, methylene chloride, diethylether, ethanol, or methanol. Mixed solvents can be used to extend the range of compounds that are effectively extracted.

Solvent extraction techniques generally suffer from the fact that relatively large volumes of solvent are required. Typically, the sample-to-solvent ratio is 1:2 or 1:3 for liquid samples. The large excess of solvent must be evaporated to concentrate the extract prior to analysis. LLE and LSE are qualitative or semiquantitative in practice.[13] Perhaps the best application for traditional solvent extraction is for the isolation of large quantities of extract from large amounts (e.g., kilograms) of sample for the purpose of trace level compound identification.

There are ways to do solvent extraction without the excessive amounts of solvent. Liquid-liquid extractors, designed to use microliter amounts of solvent, are available.[13,14] While this approach is attractive, it seems to be of limited application since recoveries are generally low.

Another approach that has proven successful for both liquid and solid samples is to use very

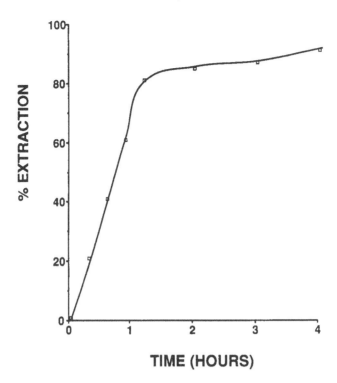

FIGURE 2.   SDE rate of extraction.

energetic mechanical agitation to mix the sample and solvent during extraction. The vigorous agitation is easily supplied by the use of a tissue homogenizer (TH). THs operate at speeds as high as 20,000 rpm. They will finely divide a solid sample and thoroughly mix the sample with the extraction solvent. Experience has shown that a sample-to-solvent ratio as high as 5:1 can be used. For example, 25 ml of orange juice (unfiltered) can be extracted with 5 ml of methylene chloride by homogenization for 30 s. The mixture is then centrifuged to separate the solid, aqueous, and solvent phases. Recoveries of the orange flavor compounds range from 60 to 90%. The TH approach does not often need additional solvent evaporation and is not as labor intensive and time consuming as traditional solvent extractions.

D. SOLID PHASE EXTRACTION (SPE)

During the last 5 years, solid phase extraction (SPE) systems have become very popular for sample preparation. The idea has been around since about 1974. Commercial products were introduced in the late 1970s. SPE is applicable to liquid samples only and can tolerate only a limited amount of particulate matter. Both volatile and nonvolatile analytes can be extracted.

Figure 4 demonstrates the characteristics of SPE systems. The extraction tubes may contain any of several high performance liquid chromatography (HPLC) packings. The tubes are prepared for extraction by rinsing with a solvent such as methanol, followed by water. When the sample is passed through the tube, the analytes of interest are quantitatively (ideally) removed from the sample matrix. The column can be washed with a *weak* solvent to remove interferences or impurities. The analytes are then eluted off the SPE column with a few milliliters of a *strong* solvent.

SPE systems are ideal for sample extraction in many respects. Degradation is minimized for thermally labile compounds. Concentration enhancements of 10- to 50-fold can be obtained

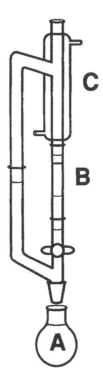

FIGURE 3. SDA apparatus. The parts are the sample flask (A), the adsorbent trap (B), and condensor (C).

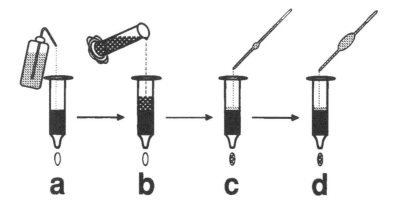

FIGURE 4. Solid phase extraction process. (a) The SPE column is conditioned by washing with methanol followed by water, (b) the sample is passed through the column to retain the analytes of interest, (c) the column can be washed with a weak solvent to remove interferences and impurities, and (d) the analytes of interest are eluted with a few ml of a strong solvent.

without additional solvent evaporation. For example, 50 ml of an aqueous sample such as a carbonated soft drink, coffee, or tea can be extracted into 2 ml of solvent. SPE sample extraction/ preparation is generally much faster than the distillation and direct solvent extraction techniques. Several samples can be processed in parallel with available vacuum manifolds. Virtually any HPLC packing material or mixtures of packings (mixtures are custom made) can be used to optimize selectivity or to increase the range of compounds extracted. Good results can be obtained with SPE columns prepared in the laboratory. RSDs of 10% or less are easy to achieve with the SPE approach.

The limitations of SPE columns can usually be worked around without much sacrifice — the main limitation being that of incomplete retention for a range of compounds with significant differences in chemical characteristics. A variety of packings may need to be evaluated or a mixture prepared to obtain a complete profile of all flavor chemicals in a sample.

Another consideration is the capacity of the SPE column. One can safely assume capacities of at least 1:10 for most compounds. For some analytes, the capacity of the SPE column is as high as 1:1, that is, 1 g of analyte can be retained on 1 g of packing material. Columns with a range of capacities (0.5 to 1 g of packing) are commercially available. If enough capacity is not obtained with available columns, custom packing or the use of more than one column in series are options.

Commercial sample preparation accessories for SPE systems use vacuum to move the sample and solvents through the columns. This is fine for nonvolatile analytes and can be used with caution for volatile analytes. The movement of volatile compounds through the SPE column is slow as it is analogous to gas-solid chromatography at room temperature. To minimize the loss of volatile compounds, a syringe can be used to apply positive pressure to move the liquids through the packing bed. References 15 to 19 provide a general overview of applications of SPE systems.

E. EXTRACT CONCENTRATION

Evaporation of the excess extraction solvent is typically the final step for the sample extraction procedures discussed so far. While it provides additional concentration enhancement, making it more feasible to analyze for and identify trace level compounds, solvent evaporation can also produce misleading results. For qualitative work, a major problem results when highly volatile compounds are totally lost during solvent evaporation. From a quantitative perspective, highly volatile compounds are more likely to be lost than are heavier ones, potentially resulting in a distortion of the relative amounts of individual components. Therefore, validation of the analysis procedure must include the effect of the solvent evaporation step. Internal standard calibration should be used to check for discrimination arising from the concentration step.

Several general points need to be kept in mind when performing solvent evaporations:

1.   The technique must be as gentle and reproducible as possible.
2.   The same amount of solvent should be evaporated from a fixed aliquot of extract.
3.   For the analysis of volatile compounds, the extract should not be allowed to go to dryness during the evaporation.

The last point is extremely critical as a drastic change in the relative amounts of the individual compounds occurs during evaporation of the last few hundred microliters of extract.

Three different basic approaches can be used to remove excess solvent:

1.   Direct blow down
2.   Indirect blow down
3.   Heated evaporation

Several variations of each approach are possible. A simple apparatus for each is shown in Figure 5.

The direct blow down approach is the most straightforward. A stream of gas, usually nitrogen, is focused onto the top of the extract through a small diameter tube such as a pipet tip, as demonstrated in Figure 5a. The sample vial may be cooled in an ice bath or other type (e.g., dry ice-acetone) of bath. Some cooling of the solution occurs from the evaporation of the solvent. Care must be taken to avoid splashing the solution out of the vial and to provide the same flow rate for each sample. This is the least reproducible and the most discriminating of the three

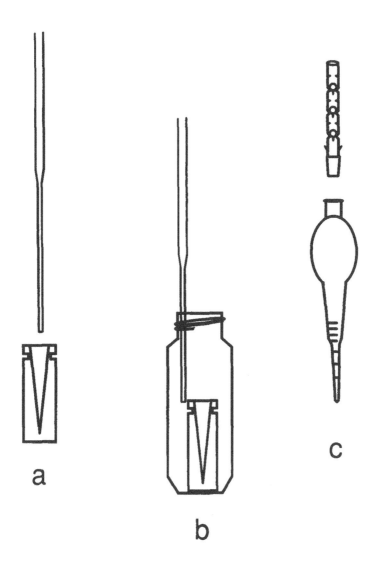

FIGURE 5. Extract concentration techniques. (a) Direct blow down, (b) indirect blow down, and (c) Kuderna-Danish apparatus.

techniques. However, it is the quickest way to remove large amounts of solvent. The direct approach is mostly applicable to nonvolatile analytes. For nonvolatile analytes, the extracts can be taken to dryness and redissolved in other solvents, such as the HPLC mobile phase, if appropriate.

A simple approach to indirect blow down is shown in Figure 5b. The stream of gas is not focused directly on the extract; rather, it sweeps the area around the vial containing the extract. This avoids some of the problems mentioned above. While somewhat slower than the direct approach, discrimination is not as much a problem. With practice and care, good results can be obtained for volatile compounds containing four or more carbon atoms. Indirect blow down represents a practical compromise for routine extract concentration.

Heated evaporation with a Kuderna-Danish apparatus shown in Figure 5c is probably the most rigorous and least discriminating of the three techniques. The temperature of the extract must be maintained at or above the boiling point of the solvent during evaporation. Deterioration of thermally labile compounds needs to be checked during procedure validation. Obviously, compounds with boiling points below that of the solvent are most likely to be lost. Depending on the amount and boiling point of the solvent, this approach may be time consuming.

F. SUPERCRITICAL FLUID EXTRACTION (SFE)

The use of supercritical fluids for extraction has been considered since the late 1800s. Interest in supercritical fluid extraction (SFE) has been cyclic and SFE has only recently been considered as a routine technique. Several different commercial extraction units are now available. Theoretically, SFE offers several potential advantages including

1.   Rapid extraction rates
2.   More efficient extractions
3.   Increased selectivity
4.   Ease of solvent removal
5.   Minimal thermal degradation
6.   Compatibility with on-line analysis

In practice, some of these advantages are difficult to realize. Extractions must often run for an hour or more to exhaustively extract a sample. Under optimized conditions, extraction efficiencies can be >95% for many compounds. Most of the work in the literature does not report quantitative recoveries. Discussion of these and many other aspects of SFE is contained in References 20 to 22.

The most common solvent for SFE is CO_2. This is because it has the lowest critical temperature and pressure, and the highest critical density of the common solvents used for SFE. Other solvents that can be used include nitrous oxide, ammonia, alcohols, tetrahydrofuran, and ethyl acetate. High critical temperatures and/or pressures make the use of these other solvents less attractive.

In practice, SFE is similar to SDE and the other exhaustive extraction techniques with respect to sample throughput and handling requirements. Extractions must run for extended periods of time requiring relatively large volumes of solvent. Compared to the other extraction techniques, solvent removal is relatively easy since CO_2 is a gas at ambient temperatures and pressures. Solute selectivity for SFE is not significantly different than that for organic solvents, that is, solutes that are more soluble are extracted most easily and completely.

SFE does seem to offer an advantage with respect to the range of chemical classes that can be extracted. The addition of small amounts of *cosolvent* or *entrainer* can significantly increase the extraction efficiencies for compounds that are difficult to extract with pure CO_2. Common entrainers are CO_2 miscible solvents such as water, ethanol, and methanol. Variations in the temperature, pressure, time, and amount of entrainer can be used to optimize extractions.

The concept of coupling on-line SFE with chromatographic techniques is interesting as it can potentially result in minimal sample handling during extraction.[21-24] As with off-line SFE, practical limitations still exist for on-line SFE, the most notable being the limited sample sizes that can be handled.

Since SFE can exhaustively extract samples with minimal degradation, some feel that it is the best way to characterize unfamiliar samples. More work needs to be done to demonstrate the real utility of SFE as a routine sample preparation tool.

G. FILTER, DILUTE, AND SHOOT (FDS)

As the title implies, it is sometimes possible to analyze liquid samples with a minimum of sample preparation. This is especially true for the analysis of nonvolatile compounds by HPLC. If the levels are high enough and interfering components do not need to be eliminated, this is an ideal way to analyze samples. Particulate matter can be removed by passing the diluted sample through a 0.45-μ membrane filter. Recoveries are effectively 100%. The precision should only be limited by the reproducibility of the instrumental technique.

This approach can sometimes be used with gas chromatography (GC) too. The main drawback is the fouling of the head of the GC column and contamination of the injection port with nonvolatile materials. Extra instrumental maintenance is required when using this technique with GC. Periodic cleaning of the injection port liner and use of a short blank retention gap column can minimize such problems.[25] Any advantages obtained by the minimal sample preparation must be weighed against the extra maintenance effort and potential expense for replacing fouled columns before this technique is chosen for use with GC.

H. HEADSPACE SAMPLING (HS)

Headspace analysis, as the name implies, is a way to monitor the composition of the gaseous space above a liquid or solid sample in a sealed container. It is best suited to the characterization of the aroma associated with a product. For example, headspace analysis would be the best way to study the aroma we experience when opening a can of coffee. Conceptually, this can be accomplished by poking a hole in the top of the can and withdrawing a sample for analysis. However, it is usually not this simple in practice.

The concentration of a given analyte in the headspace above a sample is a complex function of many variables. The composition of the headspace is a function of the air/sample partition coefficients for the volatile compounds in the sample. The partition coefficient is a measure of the relative concentration of compounds in the two phases at equilibrium as given in Equation 1:

$$K = C_{air}/C_{sample} \qquad (1)$$

C_{air} and C_{sample} are the concentrations of the analyte in the air and sample, respectively. Partition coefficients vary significantly for different compounds. They depend on several variables including temperature, concentration of the analyte in the sample, and the composition of the sample matrix (water, lipids, carbohydrates, proteins, pH, and salts).[26-28]

Rather than studying the partitioning behavior for each analyte, headspace analyses are performed by using exactly the same conditions and procedures for each sample in a series. The assumption that the effects of the above-mentioned variables are constant makes HS a practical method for routine analyses.

Many different HS schemes have been developed. Sample sizes vary from small to large volumes with either static (equilibrium) or dynamic (nonequilibrium) sampling. HS containers vary widely from larger purge and trap sizes (500 to 1000 ml) to small crimp top autosampler vials (5 to 10 ml). Since headspace analysis is used to monitor the lighter, more volatile compounds, care needs to be taken to ensure proper sample handling prior to and during sampling. All HS techniques have a few steps in common, including

1. The introduction of a fixed amount of sample into a sealable fixed-volume container
2. The equilibration at a fixed temperature for a predetermined time
3. The removal of a fixed volume of the headspace for analysis

Static sampling approximates the composition that we encounter when smelling a sample. However, the levels of sensorially active components may be so low that it is difficult to get enough sample for quantitative analysis. Static sampling is accomplished by withdrawing a sample with a gas tight syringe and introducing it into a cryogenic trap. Sampling volumes of 1 ml or more can be done with good precision by an experienced worker or automatic sampler. Typically, the RSD for static sampling is 20% or less. Addition of an internal standard can further reduce the variation. Static sampling works best for samples with relatively high levels of compounds in the headspace.

For samples with low levels of compounds, dynamic sampling, while not an equilibrium

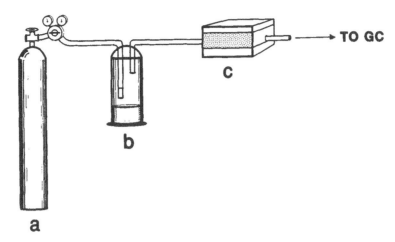

FIGURE 6.   Simplified headspace sampling apparatus. (a) carrier gas source, (b) sealed, fixed volume sample container, and (c) cryogenic or adsorbent trap.

measurement, overcomes the limited sample problem. A simplified dynamic HS scheme is shown in Figure 6. Three different dynamic sampling approaches are commonly used:

1. Sweeping the headspace above the top of the sample with a fixed volume (constant sampling time) of (carrier) gas
2. Drawing off a fixed volume (50 to 500 ml) of headspace
3. Displacing the headspace and compounds in solution by bubbling a fixed volume of (carrier) gas through the sample

The first two schemes are considered to be pseudostatic, while the last is very dynamic. During sampling the analytes are trapped on either cryogenic or adsorbent traps. Good precision can be obtained for any of these schemes when they are operated with the appropriate safeguards. Since large volumes of headspace are sampled, this approach can yield significant concentration enhancement for the analysis of trace level compounds.

Cryogenic trapping requires precise computer control and constant monitoring to ensure optimum performance. The trap can be external to the instrument or the head of the GC column. The trapped compounds are removed from external traps by controlled thermal desorption and refocused on the head of the GC column. Elaborate schemes using two traps have been reported.[29] A major problem for cryogenic traps is the formation of ice plugs for samples containing a significant amount of water. The ice plugs disrupt gas flows and lead to irreproducibility. Discrimination can occur in cryogenic traps depending on the characteristics of the individual compounds. Very volatile compounds are difficult to trap and can break through the trap during sampling. Heavy, high boiling compounds are easier to trap but may be difficult to get off the trap quantitatively. Such characteristics need to be understood prior to attempting quantitative headspace analysis with cryogenic traps.

Adsorbent trapping is analogous to solid phase extraction (SPE). Analytes are retained on the packing until they are desorbed.[7-12,30-35] Desorption of analytes can be thermal, microwave induced or with solvents. With thermal desorption, the analytes must be cryogenically focused on the head of the GC column. Cryogenic focusing is necessary to minimize the chromatographic band width. The formation of ice plugs in the cryogenic trap is not as much a problem since water can be removed from the adsorbent trap prior to desorption of the analytes by back flushing the with carrier gas. The handling of solvent desorbed samples is much like that for SPE. Functional characteristics of common trapping adsorbents are reviewed in References 8 to 12, 34, and 35. A relatively new approach to adsorbent trapping, using thick film traps, is reported in Reference 36.

III. SUMMARY

Flavor sensations, except for the four basic tastes (sweet, sour, salt, bitter) result from olefaction. Headspace analysis is the best way to compare volatile profiles, but practical and fundamental limitations can prevent it from being applied ideally. Several limitations have been described above.

Headspace procedures need to be used to measure the most volatile compounds since they are easily lost during an extraction. Determination of absolute headspace concentrations is not practical due to the difficulty in preparing accurate gaseous calibration standards. The relationship between the sample and headspace concentrations is defined by the partition coefficient for a given compound. If sample matricies are similar enough that partition coefficients are constant, the relative levels of a given compound in different samples are in the same proportion in the samples and in the headspace above the samples. Knowledge of either concentration permits calculation of the other if the partition coefficients are known. Headspace levels can be used to indirectly determine sample concentrations by calibration against matrix matched standard samples of known concentrations. Equivalent information can be obtained for many compounds in the light to medium volatility range by either headspace sampling or extraction procedures.

Properly calibrated sample extraction and analysis procedures yield the sample composition directly. They provide more information about the less volatile aroma and the nonvolatile compounds. For certain analytes and analysis procedures, sample preparation can be minimized by using the FDS approach. This is applicable mainly to the analysis of nonvolatile compounds by HPLC.

REFERENCES

1. **Reineccius, G. A.,** Determination of flavor components, in *Modern Methods of Food Analysis,* Stewart, K. and Whitaker, J., Eds., AVI Publishing, Westport, CT, 1984, chap. 12.
2. **Sugisawa, H.,** Sample preparation: isolation and concentration, in *Flavor Research — Recent Advances,* Teranishi, R., Flath, R. A., and Sugisawa, H., Eds., Marcel Dekker, New York, 1981, chap. 2.
3. **Nickerson, G. B. and Likens, S. T.,** Gas chromatographic evidence for the occurence of hop oil components in beer, *J. Chromatogr.,* 21, 1, 1966.
4. **Humbert, B. and Sandra, P.,** Determination of pyrazines in thermally treated food products: practical experiments with cocoa beans, *LC-GC,* 5, 1034, 1988.
5. **Schultz, T. H., Flath, R. A., Mon, T. R., Eggling, S. B., and Teranishi, R.,** Isolation of volatile components from a model system, *J. Agric. Food Chem.,* 25, 446, 1977.
6. **Sugisawa, H. and Hirose, T.,** Microanalysis of volatile compounds in biological materials in small quantities, in *Flavour '81,* Schrier, P., Ed., Walter de Gruyter, Berlin, 1981, 287.
7. **Ott, U. and Liardon, R.,** Automated procedure for headspace analysis by glass capillary gas chromatography, in *Flavour '81,* Schrier, P., Ed., Walter de Gruyter, Berlin, 1981, 323.
8. **Weurman, C.,** Isolation and concentration of volatiles in food odor research, *J. Agric. Food Chem.,* 17, 370, 1969.
9. **Teranishi, R., Issenberg, P., Hornstein, I. and Wick, E.,** *Flavor Research,* Marcel Dekker, New York, 1971, 37.
10. **Jennings, W. G.,** *Gas Chromatography with Glass Capillary Columns,* Academic Press, New York, 1978, 105.
11. **Sydor, R. and Pietrzyk, D. J.,** Comparison of porous copolymers and related adsorbents for the stripping of low molecular weight compounds from a flowing air stream, *Anal. Chem.,* 50, 1842, 1978.
12. **Schaefer, J.,** Comparison of adsorbents in headspace sampling, in *Flavour '81,* Schrier, P., Ed., Walter de Gruyter, Berlin, 1981, 301.
13. **Jennings, W.,** Recent developments in high resolution gas chromatography, in *Flavour '81,* Schrier, P., Ed., Walter de Gruyter, Berlin, 1981, 233.
14. **Kok, M. F., Yong, F. M., and Lim, G.,** Rapid extraction method for reproducible analysis of aroma volatiles, *J. Agric. Food Chem.,* 35, 779, 1987.

15. **Van Horne, K. C., Ed.,** *Sorbent Extraction Technology Handbook,* Analytichem International, Harbor City, CA, 1986.
16. *Waters Sep-Pak Cartridge Applications Bibliography,* Waters Chromatography Division, Millipore Corporation, Milford, MA, 1986.
17. *'Baker-10' SPE Applications Guide,* Vols. 1 and 2, J.T. Baker Chemical, Phillipsburg, NJ, 1982.
18. **Tippins, B. L.,** Solid phase extraction fundamentals, *Nature (London),* 334, 273, 1988.
19. **McDowell, R. D., Pearce, J. C., and Murkett, G. S.,** Sample preparation using bonded silica: recent experiences and new instrumentation, *TrAC,* 8, 134, 1989.
20. **McNally, M. E. P. and Wheeler, J. R.,** Increasing extraction efficiency in supercritical fluid extraction from complex matrices, *J. Chromatogr.,* 447, 53, 1988.
21. **Charpentier, B. A. and Sevenants, M. R., Eds.,** *Supercritical Fluid Extraction and Chromatography — Techniques and Applications,* American Chemical Society, Washington, D.C., 1988.
22. **King, J. W.,** Fundamentals and applications of supercritical fluid extraction in chromatographic science, *J. Chromatogr. Sci.,* 27, 355, 1989.
23. **Hawthorne, S. B., Krieger, M. S., and Miller, D. J.,** Analysis of flavor and fragance compounds using supercritical fluid extraction coupled with gas chromatography, *Anal. Chem.,* 60, 472, 1988.
24. **Andersen, M. R., Swanson, J. T., Porter, N. L.,** and Ritcher, B. E., Supercritical fluid extraction as a sample introduction method for chromatography, *J. Chromatogr. Sci.,* 27, 371, 1989.
25. **Grob, K. and Schilling, B.,** Uncoated capillary column inlets (retention gaps) in gas chromatography, *J. Chromatogr.,* 391, 3, 1987.
26. **Ioffe, B. V. and Vitenberg, A. G.,** *Headspace Analysis and Related Methods in Gas Chromatography,* John Wiley & Sons, New York, 1984.
27. **Gramshaw, J. W. and Williams, D. R.,** Physico-chemical studies on flavor active compounds, in *Flavour '81,* Schrier, P., Ed., Walter de Gruyter, Berlin, 1981, 165.
28. **Van Osnabrugge, W.,** How to flavor baked goods and snacks effectively, *Food Technol.,* 43, 74, 1989.
29. **Rodriguez, P., Eddy, C., Ridder, G., and Culbertson, C.,** Automated quartz injector/trap for fused silica capillary columns, *J. Chromatogr.,* 236, 39, 1982.
30. **Haynes, L. V. and Steimle, A. R.,** Analysis of headspace samples using off-line sorbent trapping coupled with automated thermal desorption and splitless injection onto a cryogenically cooled wide bore capillary column, *H.R.C.& C.C.,* 10, 441, 1987.
31. **Kolb, B.,** *Applied Headspace Gas Chromatography,* Heydon, London, 1980.
32. **Vercellotti, J. R., St. Angelo, A. J., Legendre, M. G., Sumrell, G., Dupuy, H. P., and Flick, G. J.,** Analysis of trace volatiles in food and beverage products involving removal at a mild temperature under vacuum, *J. Food Comp. Anal.,* 1, 239, 1988.
33. **Liardon, R. and Spadone, J. C.,** Coffee aroma characterization by combined capillary GC headspace analysis and multivariate statistics, *Colloq. Sci. Int. Cafe,(C. R.),* 11th, 181, 1985.
34. **Macleod, G. and Ames, J. M.,** Comparative assessment on the artifact background on thermal desorption of Tenax GC and Tenax TA, *J. Chromatogr.,* 355, 393, 1986.
35. **Boren, H., Grimvall, A., Palmborg, J., Savenhed, R. and Wigilius, B.,** Optimization of the open stripping system for the analysis of trace organics in water, *J. Chromatogr.,* 348, 67, 1985.
36. **Roeraade, J. and Blomberg, S.,** New methodologies in trace analysis of volatile organic compounds, *HRC & CC,* 12, 138, 1989.

GENERAL REFERENCES

1. **Maarse, H. and Belz, R.,** *Isolation, Separation and Identification of Volatile Compounds in Aroma Research,* D. Reidel, Dordrecht, The Netherlands, 1981.
2. **Zief, M. and Kiser, R.,** *Solid Phase Extraction for Sample Preparation,* J. T. Baker Chemical Co., Phillipsburg,

Chapter 3

INSTRUMENTAL CONSIDERATIONS

I. INTRODUCTION

To obtain the most information on a flavor or other complex system, the chemical or physical data used to characterize it must be related to the sensory characteristics of interest and be as detailed, precise, and accurate as possible. These three criteria are often not easy to satisfy. The more detailed the data are, the potentially more meaningful they become when trying to relate them to sensory or other product attributes. To obtain adequate chemical detail requires a high-resolution technique capable of providing both the precision and accuracy to clearly define the similarities and differences between samples.

For the analysis of both volatile and nonvolatile flavor components, modern chromatographic techniques are commonly used and are capable of meeting these criteria. In general, they can routinely provide adequate *selectivity, sensitivity, precision,* and *accuracy.* In some instances, however, they are not able to give the selectivity and sensitivity required to quantitate trace levels of very potent flavorants. One must then rely on additional semiquantitative data derived from a combination of the chromatographic technique and a more sensitive and selective detector like the human nose or mouth. For such work, the instrument is used to fractionate the sample so that its various parts can be sensorially evaluated. Some of the most meaningful information results from a combination of instrumental and human sensory responses.

For complete characterization of a system, it may be necessary to combine the data from different analytical techniques. For example, GC headspace, GC extraction, and high performance liquid chromatography (HPLC) analyses can each provide a profile of a different portion of a complex sample. The different techniques can be used to monitor a wide variety of sample characteristics. The technique(s) of choice will depend on the information to be gathered. Choosing the right analytical procedure is critical to the success of the chemometric study.

The ability of modern chromatographic techniques to provide detailed composition information for complex mixtures is primarily due to the advances in column technology over the past decade. For gas chromatography, bonded phase, fused silica capillary columns solved many of the problems and removed some of the limitations imposed by earlier column technologies. For liquid chromatography, the performance and quality of new column packings coupled with a significant increase in applications has resulted in methodologies suitable for studying complex chemical systems.

Table 1 summarizes the application areas for the most popular chromatographic techniques. Gas chromatography combined with headspace sampling (GC headspace) provides a measure of the aroma associated with a sample. The aroma is primarily composed of the lighter, more volatile compounds. The partitioning of compounds into the headspace is determined by the partition coefficient relating the concentration in the headspace and in the sample matrix as described in Chapter 2. It is inversely related to the solubility in the sample matrix and to the boiling point of the individual compounds. GC headspace analysis can be used to monitor the light-to-medium volatility compounds. Gas chromatography combined with extraction procedures extends the range of volatile compounds that can be measured. GC extraction procedures provide a picture of the heavier portion of the volatile flavor fraction. HPLC is used primarily to characterize the nonvolatile flavor fraction. In flavor terms, this would be the sweet, sour, bitter, and astringent materials. The information obtained by each technique is complementary to that obtained by the others. Each has its own set of requirements that must be met in order to provide quality chemical data.

Instrumental techniques are typically capable of better precision and accuracy than sensory evaluation techniques. For complex samples, considerable developmental effort may be re-

TABLE 1
Summary of Chromatographic Applications

Chromatographic technique	Measurement ranges			Measurement examples
	Volatility	M.W.[a]	Conc	Compounds
GC headspace	Light—medium	C_1—C_{10}	ppb—ppm	Alcohols, aldehydes esters, furans, ketones, pyrazines, phenolics, terpenoids
GC extraction	Medium—heavy	C_5—C_{30}	ppb—ppm	Same as above
HPLC	Nonvolatile	C_{10}—C_{40}	ppm—%	Acids, flavonoids, sugars, phenolics

M.W. = molecular weight.

quired to be able to analyze all the compounds over a broad range of volatilities, polarities, and chemical characteristics. The maximum amount of information can be obtained from rigorous analytical data. Rigorous implies that the recoveries for the extraction step are known, instrumental response factors have been determined for each compound, and that an acceptable calibration procedure has been used. Internal standard calibration is the method of choice as it can compensate for variations in extraction and analysis when used properly. Internal standard calibration must be used for GC analysis. For HPLC, external standard calibration can be used if variations in the extraction procedure are minimized. In the absence of absolute calibration, ratios to an internal standard can be used for comparison of similarities and differences between samples. Such ratios provide data for a given compound that are in the same proportion in different samples as the absolute amounts in those samples.

The ruggedness of an analysis procedure needs to be determined during the method development phase.[1-3] Ruggedness generally relates to the susceptibility of a procedure to variations in operator, samples, and time. Both the precision and accuracy of a procedure must be determined. Precision is a prerequisite for accuracy.[4] A precise but inaccurate (constant bias) analysis can be used for a simple comparison of several samples. However, if the data are to be used to guide flavor formulation and development work, then accuracy is also needed. References 1 to 3 provide statistical guidance, in analytical chemistry terms, that may be useful for studying and evaluating method validation data. Procedures for calculation of standard deviations (precision), determination of method bias (accuracy), linear regression, and testing for the equivalency of analytical methods are covered.

It is assumed that the reader has a working knowledge or can gain such knowledge from the references cited for the chromatographic techniques discussed below. Therefore, the discussion will be focused on the aspects of the techniques that are important for ensuring the quality of the data being gathered. A complete understanding of the characteristics of each component is necessary in order to guarantee that the data gathered will be as reproducible and accurate as possible. While the main focus is on GC and HPLC, the same concepts can be applied to other analytical measurement techniques as well.

II. CHROMATOGRAPHIC TECHNIQUES

A. GAS CHROMATOGRAPHY (GC)

GC is probably the most-used technique for the characterization of complex chemical mixtures. This is because of the amount of detailed information that can be routinely generated. Table 2 gives representative examples of the variety of analyses that can be performed by GC. An additional benefit of GC is the ease with which it can be coupled to spectroscopic and other detection schemes. A drawback is that the microgram amounts of material injected onto the

TABLE 2
Volatile Compounds Measured by Gas Chromatography

Compound class	Sensory character	Sample preparation	Examples
Aldehydes	Fruity, green	Headspace	Acetaldehyde, hexanal
	Oxidized, sweet	Extraction	Decanal, vanillin
Alcohols	Bitter, medicinal	Headspace or extraction	Linalool, menthol
	Piney, caramel	Extraction	a-Terpineol, maltol
Esters	Fruity	Headspace or extraction	Ethyl acetate, ethyl butyrate
	Citrus	Extraction	Geranial acetate
Ketones	Butter, caramel	Headspace or extraction	Diacetyl, furanones
Maillard RXN products	Brown, burnt, caramel, earthy	Headspace or extraction	Pyrazine(s), pyridine, furans
Phenolics	Medicinal, smokey	Extraction	Phenol(s), guaiacols
Terpenoids	Citrus, piney	Headspace or extraction	Limonene, pinene
	Citrus	Extraction	Valencene

column for separation are often not sufficient for collection for further characterization. A discussion of many of the relevant aspects of GC is contained in Reference 5.

The gas chromatograph is composed of four basic components. They are the injector, the column oven, the column, and the detector. Pertinent aspects of each are discussed below.

1. The Injection Port

The injection port is used to introduce the sample onto the GC column for separation and subsequent detection. Typically, microliter amounts (1 to 5 µl) of the sample extract or milliter amounts (1 to 50 ml) of headspace are injected. Split and splitless injections are the most common. Most instrument manufacturers provide injection ports that can be operated in either mode, depending on the characteristics of the sample being analyzed.

In the splitless mode, virtually all of the sample injected goes onto the head of the column.[6] The characteristics of splitless injection are summarized by Grob.[7] Splitless injection

1. Is applicable to dilute samples (0.5 to 50 ppm per compound by FID)
2. Can tolerate dirty samples
3. Produces relatively accurate results for volatile solutes
4. Can have problems with quantitation of high-boiling solutes
5. Requires reconcentration of bands by cold trapping or solvent focusing on a cool column
6. Has potential problems with retention time reproducibility

A detailed discussion of the characteristics of splitless injection is contained in References 6 to 8.

In the split mode, a fraction of the injected sample enters the column while most of the material is vented away. Split ratios of 50:1 or 100:1 (vented material:material entering the column) are common. Split injection characteristics are as follows:[7]

1. Is applicable to relatively concentrated solutions (20 ppm to 1% per compound by FID)
2. Can analyze undiluted samples
3. Is suitable for large volume headspace analysis
4. Has flexibility regarding sample concentration, injection solvent, and column temperature

5. Yields good retention time reproducibility
6. Is demanding for analysis requiring high accuracy
7. Has a risk of systematic errors

A potential problem for both split and splitless injection is thermal degradation of unstable analytes since the injection port is held at high temperatures to avoid condensation of higher boiling materials. Thermal degradation will result in absolute quantitation errors. One usually assumes that thermal degradation is not a problem unless experimental data suggest that it is occurring.

Cool, on-column injection avoids the thermal degradation problems mentioned above. It is more demanding to perform since samples are placed directly on the head of a cryogenically cooled column with special syringes. On-column injection possesses the following characteristics:[7,9]

1. Applicable to samples with a wide range of analyte concentrations (0.01 to 300 ppm)
2. Is an optimum method for producing highly accurate results
3. Requires cooling of the column to below the boiling point of the solvent
4. Is not suitable for dirty samples.

Blank retention gaps can be used in front of the analytical column to increase the injection volume and minimize the fouling of the analytical column by dirty samples.[9,10]

All three injection techniques can provide good quality data when properly used. Care must be taken to ensure that the injectors are operating properly. Periodic reanalysis of a stable sample or standard mixture of known composition is a good way to check for changes in injection efficiencies. The variation in results for reanalysis of the check sample should not be significantly greater than that obtained for successive replicate injections of the same sample with a clean injection port. If discrimination for specific compounds is seen, the injection port or column (for on-column injection) is probably fouled and needs to be cleaned. Because of the potential for variation in the amounts of sample reaching the column, internal standard calibration should be used. When operating properly, the precision of absolute peak areas, as measured by the relative standard deviation (RSD) for replicate injections of the same sample, should be less than 10%, and the RSD based on internal standard calibration should be between 5 and 10%.

The injection technique of choice will depend on the characteristics of the samples to be analyzed and the availability of instrumentation. Considering the number of analyses to be performed for collection of the chemical database and the need for high quality data, techniques that can be automated should be considered first.

2. The Column Oven

Virtually all instrument manufacturers offer options that allow precise and accurate control of the oven temperature, from cryogenic (usually liquid N_2) to over 400°C. The range of temperatures required for a particular analysis will depend on the sample composition, injection technique, and the thermal limits of the column. For relatively simple samples containing only a few compounds, isothermal analysis may be possible and offers the advantages of simplicity and good reproducibility. However, most samples will be sufficiently complex that temperature programming of the oven will be required to adequately resolve a large number of components with a wide range of boiling points. Since modern instruments are designed to operate either isothermally or in a temperature-programmed mode, either can be easily implemented as needed.

Headspace analysis requires cryogenic cooling during the injection to minimize the sample band width, thereby permitting the resolution of highly volatile compounds. Analysis of solvent extracts, on the other hand, may not require initial cryogenic conditions since the practical analysis of low-boiling compounds is limited by the boiling point of the extraction/injection solvent.

TABLE 3
Modern Capillary GC Column Characteristics

Polarity	Stationary phase composition	Structure	Temp range (°C)	Bonded phase equivalents	Nonbonded phase equivalents
Nonpolar	100% dimethyl-polysiloxane	$\begin{bmatrix} CH_3 \\ \mid \\ -Si-O- \\ \mid \\ CH_3 \end{bmatrix}_n$	−60—300	DB-1, SPB-1, MS, CPSIL 5CB	OV, OV-101, SE-30
	95% dimethyl-5%-5%-diphenyl-polysiloxane	$\begin{bmatrix} Ph \\ \mid \\ -Si-O- \\ \mid \\ Ph \end{bmatrix}_n \begin{bmatrix} CH_3 \\ \mid \\ -Si-O- \\ \mid \\ CH_3 \end{bmatrix}_n$	−60—300	DB-5, SPB-5, MPS-5, CPSIL 8CB	OV-73, SE-52, SE-54
Intermediate	100% methyl-phenylpolysilo-xane	$\begin{bmatrix} CH_3 \\ \mid \\ -Si-O- \\ \mid \\ Ph \end{bmatrix}_n$	25—300	DB-17, SPB-2250, MPS-50, CPSIL 19CB	OV-17, OV-1701
	86% dimethyl-14%-cyano-propylphenyl-polysiloxane	$\begin{bmatrix} CH_3 \\ \mid \\ -Si-O- \\ \mid \\ CH_3 \end{bmatrix}_n \begin{bmatrix} Ph \\ \mid \\ -Si-O- \\ \mid \\ C_3H_6 \\ \mid \\ CN \end{bmatrix}_n$	25—300	DB-1701, SPB35, MPS-50, CPSIL 19CB	OV-17, OV-1701
Polar	50% cyanopropyl-methyl-50% methylphenyl-polysiloxane	$\begin{bmatrix} CH_3 \;\; CH_3 \\ \mid \quad\;\; \mid \\ -Si-O-Si-O- \\ \mid \quad\;\;\; \mid \\ C_3H_6 \;\; Ph \\ \mid \\ CN \end{bmatrix}_n$	40—240	DB-225, CPSIL 84	OV-225
	Polyethylene glycol	$[H-O-CH_2-CH_2-]_n-OH$	20—260	DBWAX, SUPELCOWAX, CPWAX	CARBOWAX-20M

3. The Analytical Column

Most studies where chemometric interpretation of data is required involve samples of sufficient complexity that capillary columns are needed. Fused silica capillary column technology has undergone significant improvements during the last decade. Commercial columns are available that are easy to work with (durable and flexible) and have excellent thermal stability (bonded phases). A wide variety of stationary phases (both bonded and coated) and capacities coupled with high efficiencies (up to 4000 theoretical plates per meter) allow the simultaneous analysis of literally hundreds of compounds in a single sample. Table 3 summarizes the characteristics of a variety of columns that are commercially available. This table is not all inclusive, but demonstrates the range of characteristics available.

Columns with the stationary phase covalently cross-linked and bonded to the walls of the column provide superior stability compared to coated columns. Bonding minimizes bleeding, allows thicker phases for greater capacity, and extends the working temperature range of the column. Additionally, bonded phase columns can be backwashed with solvent to remove impurities to restore column performance.

Column selection depends on the analytes to be quantitated. Most columns have similar thermal ranges and capacities. Polarity of the stationary phase is the most important parameter

TABLE 4
Common Gas Chromatographic Detectors

Detector	Applications	LLD[a]	LDR[b]	Selectivity	Other comments[c]
Flame ionization (FID)	Most organic compounds	10^{-10}	10^7	<10	Type 2; general purpose detector
Flame photometric (FPD)	Organic sulfur, phosphorus compounds	10^{-11}	10^3	>10^5	Type 2; limited applications
Thermionic emission (TID)	Organic sulfur, phosphorus, nitrogen, and halogen compounds	10^{-12}	10^3	<10^4	Type 2; operating conditions determine specificity
Electron capture (ECD)	Halogenated, poly-aromatic compounds	10^{-14}	10^2	Compound dependent	Type 1; radioactive source
Thermal conductivity (TCD)	Universal	10^{-7}	10^4	<10	Type 1; first GC detector; limited use with capillary columns
Atomic emission detector (AED)	Universal or specific	10^{-12}	10^3	10^3—10^5; compound dependent	Type 2; specificity determined by wavelength
Mass spectrometry (MS)	Universal or specific	10^{-8}	10^3	Compound dependent	Best detection limits and quantitation by selected ion monitoring
Fourier transform infrared spectroscopy (FTIR)	Universal or specific	10^{-7}	10^3	Compound dependent	Best detection limits and quantitaton by selected wavelength monitoring

[a]   Practical lower limit of detection; type 1 — mol/cc; type 2 — mol/s.
[b]   Linear dynamic range.
[c]   See text for discussion of detector types.

for column selection. The polarity of the column should be matched to the characteristics of the analytes when possible. Samples containing compounds with a wide a range of chemical characteristics can usually be analyzed on a column of intermediate polarity. Although not common, it is possible to use two columns of different polarity in series to optimize separations. Agricultural, analytical chemistry, environmental, food science, and manufacturers' literature are full of examples for analysis of a wide variety of complex samples. A search of these areas will help in the selection of the appropriate column.

4. The Detector

GC detectors generally fall into two categories: those that respond to the concentration (mole fraction) of the analyte in the gas stream (type 1) and those that respond to the mass flow rate (moles per second) of the analyte (type 2).[11] The differences result from the fundamental modes of detector operation. Type 1 detectors detect the presence of an analyte by measuring a change in a bulk property of the gaseous media. These detectors usually are not destructive, permitting collection of fractions for further characterization. However, the *detection limits, linear dynamic range, sensitivity*, and operating conditions limit the applicability of these detectors. Type 2 detectors are destructive because they operate by decomposing the analyte in a flame, followed by detection of the reaction products. The integrated signal is proportional to the total mass of the analyte present.

Table 4 summarizes the characteristics of the most common GC detectors. References 11 to 13 provide detailed information about many GC detectors.

The flame ionization detector (FID) is the most universally applicable detector for GC. It provides a response to any compound that is combustible in a hydrogen/air flame. The FID is destructive and thus prevents further characterization of analytes after detection. The wide linear dynamic range and good sensitivity make it the detector of choice for most general applications.

Many other detectors are available for GC. Except for the thermal conductivity detector (TCD) they are generally more selective and do not have the linear dynamic range of the FID. Element-selective detectors are available for halogens, nitrogen, sulfur, and phosphorus in organic compounds. Their sensitivity is equal to or greater than the FID for the selected compounds. Saturation at higher levels results in nonlinear response and determines the upper limit of the linear dynamic range. Research continues on the development of selective detectors with improved charactersitics.[12-14]

The plasma atomic emission detector (AED) provides the specific-element detection capabilities of the other selective detectors for several elements. Si, N, P, S, C, H, O, and halogens can be detected. Depending on instrumental configuration, multiple elements can be monitored simultaneously. Choice of C and H emission wavelengths make it a universal detector for organic compounds, while selectivity is obtained by choosing the wavelengths appropriate for the specific elements of interest. Many research and development applications of this type of detector are reported in the literature. The actual utility of this detector as a routine tool is yet to be determined, as a commercial unit has only been available for about 1 year.

All the detectors mentioned above are designed for quantitative analysis. GC can be coupled with spectroscopic detectors, such as Fourier transform infrared spectrometers (GC-FTIR) and mass spectrometers (GC-MS), for identification of individual compounds. These detectors can also be used for selective monitoring of compounds containing specific functional characteristics. For example, selected wavelengths can be monitored in the IR for analysis of compounds containing specific functional groups. Selected ions can be monitored in the mass spectrum to detect compounds with certain structural characteristics. Quantitation can be performed in the selected monitoring modes for these detectors. Experts in the respective areas should be consulted to determine if these techniques can be used in the particular application.

The last detector to be discussed is the human nose. Nasal detection is similar to instrumental detectors in some respects and different in others. It is similar in that the sensitivity and selectivity are compound dependent, and the response can depend on the chromatographic resolution of individual compounds. It is different in that quantitation of response is less precise and typically nonlinear. For certain compounds, the nose is much more sensitive than any instrumental detector. A trained nose is a very powerful tool for the identification of sensorially active compounds.

The GC *sniffport* technique is not new.[15,16] It has been used for many years. Conceptually, it is simple — substitute the nose for an instrumental detector. However, proper implementation requires attention to design details and operator training to obtain good results. Figure 1 demonstrates the sniffport technique. Splitting the column effluent between an instrumental detector and a sniffport allows simultaneous detection and makes relating the two sets of data easier. Like instrumental detectors, makeup gas is needed to keep the gas stream moving and to condition it for detection. For the sniffport, moist air is a good makeup gas as it helps to minimize fatigue caused by drying of the nasal membranes by the hot column effluent. The sniffport base should be heated (like instrumental detectors) to avoid condensation on cool surfaces. The distance between the end of the GC column (in the heated base) and the point where the detection occurs must be kept to a minimum to avoid condensation. These conditions can easily be met through the use of commercially available or custom manufactured parts.

Compounds detected by the nose are likely to be important to a food or beverage flavor. However, focusing on the those compounds only can be misleading. The sensory response to a given compound eluting from the GC column depends on the amount of that compound present in the extract, the amount of the extract injected onto the column, the plumbing between the

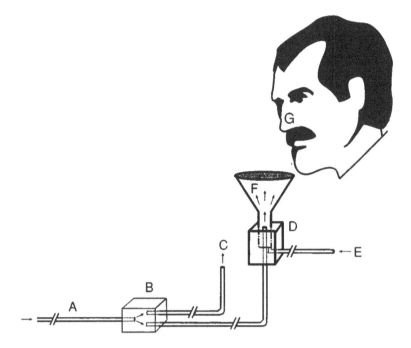

FIGURE 1.   Schematic of GC sniffport. Effleunt from the GC column (A) enters the splitter (B), part going to the instrumental detector (C) and part going to the sniffport. The sniffport is designed similar to instrumental detectors in that a heated base (D) and makeup gas (E) are required. The effluent from the sniffport is focused by a funnel or cup (F) for nasal (G) evaluation. See text for detailed discussion.

column exit and the sniffport, and the sensitivity of the person doing the sniffing. Odor detection thresholds can vary significantly from person to person and from day to day for the same person. Additionally, some compounds that are nonodorous in the GC effluent can produce an effect when mixed with other compounds. A good example of this is the pyrazines commonly found in thermally processed foods. Many have relatively high odor thresholds and cannot be detected individually at the sniffport. In combination they provide part of the characteristic aroma of products like coffee, chocolate, and baked goods.

An additional benefit of the sniffport is that the column effluent can be trapped, without major instrumental changes, for the evaluation of the complete aroma character.[17] If the trapped material faithfully reproduces the original sample, the extraction and analysis procedures have not biased the instrumental measurements. If a significant distortion is noted, then the entire analysis procedure needs to be reevaluated.

For all the benefits of great sensitivity and selectivity, being able to smell compounds eluting from the GC when they cannot be detected instrumentally clearly shows the limitations of current instrumental detection techniques. This situation is both challenging and frustrating for the scientist gathering the chemical data. Ultimately some information describing the relative magnitudes of trace compounds needs to be developed to complete the sample characterization. The combination of instrumental and nasal detection is currently the best approach to complex flavor analysis, although the data must be considered semiquantitative at best. Many of the same procedures that are used for sensory data collection, such as threshold determination, category scaling, or magnitude estimation, can be used for collection of data at the sniffport.

B. HIGH PERFORMANCE LIQUID CHROMATOGRAPHY (HPLC)

HPLC is gaining popularity as a means of characterizing the nonvolatile chemical composition of complex samples. HPLC complements the information obtained by GC. The com-

TABLE 5
Flavor Compounds Analyzed by HPLC

Compound class	Sensory character	Examples	Sample preparation	Detection
Acids				
Amino acids	Sweet, sour, bitter	All known	Derivatize	UV @254 nm
Organic acids	Sour	Citric, malic	FDS	UV @210 nm
Polyphenolic acids	Astringent, bitter	Chlorogenic	FDS	UV @320 nm EC
Flavonoids	Astringent, bitter	Flavonols, anthocyanins	SPE	UV-VIS EC
Phenolics	Medicinal, smokey	Guaiacols, phenols	FDS, SPE	UV @280 nm EC
Sweeteners				
Sugars	Sweet, body	Sucrose, glucose, fructose	FDS, SPE	RI; UV @210 nm
Artificial	Sweet	Aspartame	FDS, SPE	UV @254nm

bination of these two techniques yields more detailed sample composition information than ever before. Most of the reason for increased application of HPLC is improvements in column technology. The development of bonded phase column packings has made analyses possible that were difficult or even impossible 10 years ago. Moderately priced, high-efficiency columns are available that allow the simultaneous analysis of many compounds. For example, detailed information about organic acids and sugar profiles can be used instead of the less-specific measurements of titratable acidity and °brix, respectively. Symposia have been organized and books have been written specifically on HPLC applications to foods and beverages.[18,19] Classes of compounds, previously not well characterized, are now being studied by HPLC techniques.[20-22] Reference 23 provides an overview of HPLC applications to foods and beverages. Table 5 contains examples of the classes of chemicals in foods and beverages that can be routinely monitored by HPLC.

An advantage of HPLC is the ease with which fractions can be collected for further characterization. Combinations of preparative and analytical scale HPLC can be used to obtain the maximum purification of individual compounds for characterization.[24] Instrumental, chemical, or sensory testing of the isolated fractions is possible. Analytical instruments can be used to develop separation schemes and isolate small amounts of material. For larger quantities (up to grams), semipreparative or preparative scale systems are required. Ideally, analytical separation conditions can be applied to the preparative separations. In practice, the reduced resolution in the preparative systems and the potential toxicity of many solvents prevent the translation between analytical and preparative work. If the collected fractions are for sensory evaluation (tasting), the choice of usable solvents is limited. Ethanol is the solvent of choice in this situation. Since ethanol is not normally used for analytical separations, additional development effort is required when converting from an analytical to preparative scale separation.

HPLC instrumentation is composed of a solvent delivery system [pump(s)], an injector, a column, and a detector. While different designs of each are available, they all need the same functional characteristics. Pertinent aspects of each are discussed below. More detailed discussions of specific design considerations are contained in References 25 and 26.

1. The Solvent Delivery System

The solvent delivery system or pump needs to provide a reproducible and constant flow of solvent to the rest of the instrument. Most are designed to provide constant flow at any operating pressure. Almost all commercially available solvent delivery systems are capable of meeting the criteria of reproducibly delivering solvent and thus providing constant retention times.

Either *isocratic* (constant solvent composition) or *gradient* (varying solvent composition) elution can be used. Most complex analyses require the use of gradient conditions to obtain the maximum separation for solutes with a wide variation in characteristics. Binary gradients are the most common. Ternary gradients can be formed on some instruments. After a gradient elution, the system must be returned to initial conditions and the column reequilibrated. Small changes in the gradient solvent composition and inadequate column equilibration can result in retention time irreproducibility. For these reasons, less-complex solvent systems are preferred for routine work.

2. Sample Injectors

Fixed-volume loop-style injectors are the most common and the most reliable. They are capable of precisely introducing the same amount of sample each time. Because of the excellent precision obtained with this style of injector, external standard calibration can be used. Nonfixed-volume loop injectors are less common and are sometimes incorporated into autosamplers, thus allowing one to change injection volumes without having to change the sample loop. The same volume is delivered to the instrument by a computer-controlled syringe. The syringe draws a predetermined amount of sample into an oversized loop. While not as foolproof as the fixed-volume loop injectors, this design is capable of good precision when operating properly. External standard calibration is possible with this style of injector as well.

Since the volume of sample injected into the HPLC ranges from 10 to 100 µl (or more), the error caused by minor variations in injection volume has less overall effect than the same absolute error in GC. That is, an absolute error of 0.1 µl in injection volume is a 10% error for 1 µl injected into a GC, but is a 1% error for a 10 µl and a 0.1% error for a 100-µl HPLC injection. These smaller relative errors combined with ambient injection temperatures make precise (<1% RSD) injection in HPLC easy to accomplish.

3. The Analysis Columns

Bonded stationary phases are used almost exclusively for food and beverage analyses, and offer several advantages over traditional stationary phases. Bonded phases are created by covalently linking selected functionalities to the silica or porous polymer supports. Since the stationary phase is chemically bonded to the support, it is not easily removed during analysis, thus yielding more stable columns. A variety of functional characteristics provides the needed flexibility for analysis of most common solutes. Table 6 summarizes some of the most common stationary phases and typical applications for each. As in GC, the polarity of the stationary phase needs to be matched to that of the analytes. Unlike GC, the HPLC mobile phase must also be optimized to yield the desired separation, depending on the stationary phase characteristics.

The exact retention mechanisms are not known for bonded stationary phases. They are usually considered to be equivalent to mechanically held liquid phases. This assumption leads to reasonable predictions of retention as a function of solute structure and mobile phase composition.[26] Other considerations include specific chemical interactions and/or size interactions between the stationary phase and solutes.

Normal phase refers to chromatography with relatively polar stationary phases and relatively nonpolar mobile phases. The packings can be underivatized silica where adsorption is the primary separation mechanism or polar, chemically bonded phases where ion exchange and other mechanisms are operating. Polar solutes can be separated by the use of non-polar solvents or mixtures of solvents. Normal phase separation is applied mainly to the analysis of vitamins, sugars, acids, and bases from food and beverage products.

Reverse phase implies that the mobile phase is more polar than the stationary phase, just the opposite of normal phase. With reverse phase packings, mixed aqueous/organic mobile phases are used. The strength of the mobile phase is determined by the amount of water-miscible organic solvent, the pH, or the ionic strength. Variations in these characteristics are used to form gradients for reverse phase HPLC.

TABLE 6
Modern Bonded Stationary Phase HPLC Packing Materials

Polarity	Name	Active functionality	Separation mechanism
Nonpolar	Reverse	C_1, C_8, C_{18} alkylsilane	Partitioning
	Phenyl	Phenyl	Partitioning
Intermediate	Cyano	Nitrile	Partitioning, adsorption[a]
	Amino	Amine	Partitioning, adsorption, or ion exchange[a]
Polar	Normal	Silica[b]	Adsorption
	Ion exchange	SO_3^-; NR_3^+	Anion; cation exchange

[a] Depends on separation conditions.
[b] Not derivatized.

Reverse phase packings vary in the amount of bonded organic molecules (called carbon loading) and the character of the bonded compounds. Depending on the carbon loading (5 to 15%), the degree of residual silanol deactivation, and the solvent system, the separation mechanism can be partitioning (between stationary and mobile phase), adsorption, and/or ion exchange. For example, for a polar cyanoamino stationary phase, three different modes of operation can be used depending on the solvent composition. With a polar solvent, the stationary phase will act as a reverse phase of moderate polarity. An acidic aqueous solvent produces a weak anion exchange mechanism, while an acidic nonaqueous solvent results in ion pairing mechanisms. The wide variety of stationary phases and convenience has made reverse phase the HPLC method of choice for most applications. The manufacturer's application notes and the literature should be consulted to identify the combination of stationary phase and mobile phase best suited for a given application.

Both aqueous and nonaqueous size exclusion (SEC) or gel permeation (GPC) columns are available in analytical and preparative sizes. The ability to work with aqueous samples/extracts has greatly expanded the utility of SEC. However, due to the inherent limitations of the separation mechanism, SEC is not generally applicable. Most flavor molecules are relatively small (<500 molecular weight) and will all tend to elute under one broad peak. SEC systems are useful for the characterization of polymeric materials (polysaccharides and polyphenols) with a broad range in molecular weight distribution.

The inexperienced user needs to be aware of some practical limitations of column technology. Not all reverse phases, although nominally the same, are equivalent. This is due to variations in derivatization chemistry and silica deactivation procedures. Manufacturer-to-manufacturer variations are larger than batch-to-batch variations for the same manufacturer. Silica-based stationary phases have a working pH range of 2 to 8, whereas the polymer based phases can be used between pH 1 and 13. Stationary phase particle size influences the chromatographic efficiency (i.e., resolution), as well as the system operating pressure which can limit the solvent flow rate. Smaller particles, 3 or 5 μ in diameter, yield more efficiency and higher operating pressures at the same flow rate than the larger particles (10 μ). Routine procedures can be implemented on either 5- or 10-μ particle size columns.

Column capacity refers to the amount of analyte that can be chromatographed without loss of efficiency. Overloading a column can significantly reduce or eliminate the resolution between adjacent peaks. The capacity per compound ranges from microgram quantities for microbore columns to gram quantities for preparative columns. Analytical columns can tolerate a wide range of material from micrograms to milligrams. Because of the limited capacities and the requirements for specialized solvent delivery systems and detectors, packed microbore (1 mm diameter) and capillary (<1 mm diameter) columns are not routinely used. These practical

TABLE 7
HPLC Detectors

Detector	Applications	LDR[a]	Specificity	Gradient elution
UV-VIS	Compounds with chromophores	10^5	Determined by wavelength	Yes
EC	Compounds with electroactive groups	10^6	Potential dependent	No
RI	Universal	10^4	None	No
IR	Universal, identification	10^4	Determined by wavelength	Yes
MS	Universal, identification	10^4	Determined by ions selected	Yes
Fluorimeter	Fluorescing compounds	10^3	Determined by excitation and emission wavelengths	Yes
Radioactivity	^{14}C-containing compounds	10^4	^{14}C	Yes

[a] Linear dynamic range.

limitations outweigh the advantages of low solvent usage, high efficiency, and high mass sensitivity offered by such systems. At the other extreme, the limited efficiency and high solvent usage of semipreparative and preparative columns limit their routine utility for quantitative analysis.

Fouling of columns, by material retained from the sample, can be minimized by use of disposable guard columns. The guard column can contain any packing material, but usually is similar to the analytical column. Guard columns do not add to or decrease the efficiency of the overall system, but significantly extend the life of the more expensive analytical columns.

4. The Detector

Like GC, many different types of detectors are available for HPLC. The same characteristics are important for both GC and HPLC. They include response to the compounds of interest, linear dynamic range, and sensitivity. Most HPLC detectors respond to the concentration of the analyte. As in GC, both universal and specific detectors are available. Table 7 summarizes the characteristics of the most common HPLC detectors.

The refractive index (RI) detector is the most universal HPLC detector. Practical limitations such as low sensitivity, time- and temperature-induced drift, and incompatibility for use with gradient solvent programming greatly restrict areas of RI application. It is often used for analysis of sugars and simple organic acids, since these molecules are difficult to detect by other means and are usually present at relatively high levels (%) in food and beverage systems.

The ultraviolet-visible (UV-VIS) detector is the most common HPLC detector. This is because of its good sensitivity, wide linear dynamic range, stability, and compatibility with gradient solvent programming. Variable wavelength detectors, while not as sensitive as fixed wavelength detectors, provide an easy way of selectively monitoring different classes of compounds depending on their UV-VIS absorption characteristics. Diode array detectors provide the most information, since data over a selected wavelength range can be obtained for each peak during an analysis. Diode array detectors generate large amounts of data during an HPLC run. Megabytes of information are obtained when a complete spectrum is obtained every few seconds. Therefore, they are often used during methods development for selection of the wavelength(s) to monitor. Routine sample analyses are then performed by monitoring the selected wavelengths only.

Several types of selective detectors are also available for HPLC. These are designed to monitor compounds with certain chemical characteristics, as opposed to those containing specific elements. The most common are electrochemical (EC) (amperometric), fluorescence, conductivity, and radioactivity. As the names imply, each takes advantage of a particular chemical characteristic for detection. They offer good sensitivity to selected chemicals and good linear dynamic ranges. The fluorescence and radioactivity detectors are stable with respect to time and temperature, and are compatible with gradient solvent programming. References 26 and 27 discuss these and other detectors in detail.

Spectroscopic detectors, FTIR, and MS can also be used with HPLC. The combination of these techniques with HPLC is more recent than for GC. They should offer all capabilities for HPLC that they do for GC.

C. MULTIDIMENSIONAL CHROMATOGRAPHIC TECHNIQUES

Work is currently focused on the coupling of chromatographic techniques to enhance the resolution and separation of trace level compounds. The idea of combining multiple steps in a fractionation/analysis sequence is not new. For years workers have used sequential procedures to isolate fractions of successively greater purity. Most of the interfaces between the sequential steps were manual. It has recently become possible to combine multiple chromatographic procedures into totally automated instrumentation. These techniques are currently being applied to isolation and qualitative identification of compounds in complex mixtures.

All possible combinations of multidimensional chromatographic techniques are being explored. LC-LC is the most straightforward of the techniques. It can be implemented easily with standard instrumental components. The combination of HPLC with capillary GC is relatively new.[34,35] Instruments to perform two-dimensional GC are available commercially.[36,37] Reference 34 provides an overview of both instrumentation and applications. Reference 36 describes the combination of many of the sampling, preparation, and analysis schemes described in Chapters 2 and 3.

The basis of multidimensional techniques is called *heart cutting*. It basically consists of selecting a limited portion of the first chromatogram and sending it to the second chromatographic column for more complete separation of the individual compounds. This idea is demonstrated in Figure 2. Frequently a single peak in the first chromatogram is separated into multiple peaks in the second chromatogram. This allows the more exact identification of the specific compound responsible for the sniffport response in GC analysis or the taste obtained in HPLC fractions. These techniques are very powerful for isolation and identification. However, the complexity of the output that would be obtained by heart cutting every portion of the first chromatogram prevents practical application for collection of data for chemometric analysis at this point in time.

III. SPECTROSCOPIC TECHNIQUES

Spectroscopic techniques such FTIR and UV-VIS offer the ability to monitor certain characteristics in bulk samples of foods and beverages without chromatographic separations. With these techniques, one can rapidly analyze samples with minimal sample preparation. However, they generally do not provide a complete profile, as many of the volatile aroma compounds are present at levels too low for the FTIR to detect or do not absorb UV-VIS radiation. The spectroscopic techniques can be used to monitor selected components. Since there is a high degree of correlation between the spectral intensities at the different wavelengths, continuous modeling techniques, such as principal components analysis (PCA) and partial least squares (PLS) use their averaging effect to great benefit and allow the extraction of useful information. It is virtually impossible to extract such information without the use of multivariate data analysis techniques. References 38 to 40 describe some spectroscopic applications.

For just the ability to predict characteristics without the desire or need to understand the cause of the effect, these techniques are potentially very useful. However, to search for cause and effect relationships, some chromatographic work will probably be required. Interpretation of the relationships between composition and sensory character is not as straightforward with the spectroscopic approaches. The identity of the compounds yielding the spectral information that

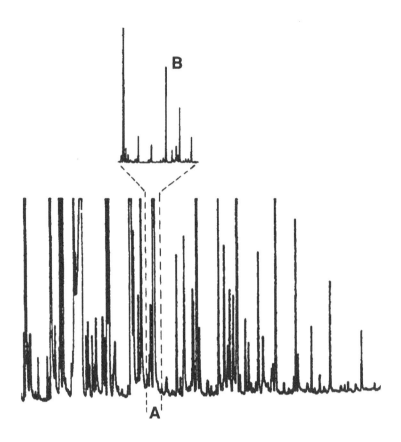

FIGURE 2.   Two-dimensional heart cutting technique. A selected region (A) from the
first column is switched to the second column (B) where additional separation occurs.

leads to the mathematical model is not always obvious. An interpretative approach must be used
to determine the characteristics of the compounds of interest. Different wavelengths will be
indicative of certain functionalities or chromophoric groups in molecules which can often be
related to specific classes of compounds. Finding the interrelationships between the spectroscopic data and the chromatographic data is necessary. Multivariate analysis techniques can be
used to compare the samples with each data set and to look for similar trends. When the
compounds are found that are responsible for the observed spectral characteristics, the
interrelationships between the samples will be the same for both data sets.

IV. OTHER TECHNIQUES

The above detailed discussions focused primarily on GC and HPLC because of the utility they
offer for studying flavor and other complex systems. Many other types of instrumental or
chemical data can be gathered. For example, supercritical fluid chromatography (SFC),[28,29]
texture measurements,[30,31] and wet chemical data[32,33] can all be related to sensory data as well.
The same quality of data is required, and most of the principles of sample handling, preparation,
and measurement apply no matter which measurement technique is used for physical or
chemical characterization.

REFERENCES

1. **Mulholland, M.,** Ruggedness testing in analytical chemistry, *TrAC*, 7, 383, 1988.
2. **Caulcutt, R. and Boddy, R.,** *Statistics for Analytical Chemists*, Chapman and Hall, New York, 1983.
3. **Massart, D. L., Vandeginste, B. G. M., Deming, S. N., Michotte, Y., and Kaufman, L.,** *Chemometrics: A Textbook*, Elsevier, Amsterdam, 1988.
4. **Bevington, P. R.,** *Data Reduction and Error Analysis for the Physical Sciences*, McGraw-Hill, New York, 1969, 4.
5. **Jennings, W.,** Gas chromatography, in *Modern Methods of Food Analysis*, Stewart, K. and Whitaker, J., Eds., AVI Publishing, Westport, CT, 1984, 319.
6. **Grob, K., Biedermann, M., and Li, Z.,** Checking the capacity of a splitless injector — a simple test, *J. Chromatogr.*, 448, 387, 1988.
7. **Grob, K.,** *Classical Split and Splitless Injection in Capillary GC*, Huethig, Heidelberg, 1986.
8. **Grob. K., Laubli, Th., and Brechbuhler, B.,** Splitless injection — development and state of the art, *H.R.C. & C.C.*, 11, 462, 1988.
9. **Grob, K.,** *On-Column Injection in Capillary Gas Chromatography*, Huethig, Heildelberg, 1987.
10. **Grob, K. and Schilling, B.,** Uncoated capillary column inlets (retention gaps) in gas chromatography, *J. Chromatogr.*, 391, 3, 1987.
11. **David, D. J.,** *Gas Chromatographic Detectors*, John Wiley & Sons, New York, 1974.
12. **Dressler, M.,** *Selective Gas Chromatographic Detectors*, Elsevier, Amsterdam, 1986.
13. **Patterson, P. L.,** Recent advances in thermionic ionization detection for gas chromatography, *J. Chromatogr. Sci.*, 24, 41, 1986.
14. **Driscoll, J. N. and Berger, A. W.,** Improved flame photometric detector for the analysis of sulfur compounds by gas chromatography, *J. Chromatogr.*, 468, 303, 1989.
15. **Fuller, G. H., Steltenkamp, R., and Tisserand, G. A.,** The gas chromatograph with human sensor: perfumer model, *Ann. N.Y. Acad. Sci.*, 116, 711, 1964.
16. **Acree, T. E., Barnard, J., and Cunnignham, D. G.,** A procedure for the sensory analysis of gas chromatographic effluents, *Food Chem.*, 14, 273, 1984.
17. **Adam, S. T.,** A simple and efficient cryogenic trapping device for odor evaluation of aroma fractions, *H.R.C. & C.C.*, 10, 369, 1987.
18. **Charalambous, G., Ed.,** *Liquid Chromatographic Analysis of Food and Beverages*, Vol. 2, Academic Press, New York, 1979.
19. **Kirk, J. R.,** Modern liquid chromatography: evolution and benefits, in *Modern Methods of Food Analysis*, Stewart, K. and Whitaker, J., Eds., AVI Publishing, Westport, CT, 1984, 381.
20. **Lunte, S. M., Blankenship, K. D., and Read S. A.,** Detection and identification of procyanidins and flavonols in wine by dual-electrode liquid chromatography-electrochemistry, *Analyst (London)*, 113, 99, 1988.
21. **Daigle, D. J. and Conkerton, E .J.,** Analysis of flavonoids by HPLC: an update, *J. Liq. Chromatogr.*, 11, 309, 1988.
22. **Delcour, J. A., Vinkx, C. J. A., Vanhamel, S., and Block, G. G. A. G.,** Combined monitoring of UV absorbance and flourescence intensity as a diagnostic criterion in reversed-phase high performance liquid chromatography separations of natural phenolic acids, *J. Chromatogr.*, 467, 149, 1989.
23. **Martin-Hernandez, M. C. and Juarez, M.,** Chromatographic techniques in food analysis, *Alimentaria*, 24, 13, 1987.
24. **Gutierrez, F., Albi, M. A., Palma, R., Rios, J. J., and Olias, J. M.,** Bitter taste of virgin olive oil: correlation of sensory evaluation and instrumental HPLC analysis, *J. Food Sci.*, 54, 68, 1989.
25. **Howard, P. Y. and Hodgin, J. C.,** A guide to HPLC instrument selection in the food science laboratory, in *Liquid Chromatographic Analysis of Food and Beverages*, Vol. 2, Charalambous, G., Ed., Academic Press, New York, 1979, 255.
26. **Snyder, L. R. and Kirkland J. J.,** *Introduction to Modern Liquid Chromatography*, 2nd ed., John Wiley & Sons, New York, 1979.
27. **Scott, R. P. W.,** *Liquid Chromatography Detectors*, Elsevier, New York, 1977.
28. **Charpentier, B. A. and Sevenants, M. R., Eds.,** *Supercritical Fluid Extraction and Chromatography — Techniques and Applications*, American Chemical Society, Washington, D.C., 1988.
29. **Wheeler, J. R. and McNally, M .E.,** Is SFC worth the effort?, *Res. Dev.*, 31, 134, 1989.
30. **Vickers, Z. M.,** Instrumental measures of crispness and their correlation with sensory assessment, *J. Texture Stud.*, 19, 1, 1988.
31. **Baruch, D. W. and Atkins, T. D.,** Using the wheat research institute chomper to assess crumb flexibility of staling bread, *Cereal Chem.*, 66, 59, 1989.
32. **Martens, M., Fjeldsenden, B., Russwurm, H., Jr., and Martens, H.,** Relationships between sensory and chemical quality criteria for carrots studied by multivariate data analysis, in *Sensory Quality in Foods and Beverages*, Williams, A. A. and Atkin, R. K., Eds., Ellis Horwood, Chichester, England, 1983, 233.

33. **Martens, M. and Van der Burg, E.,** Relating sensory and instrumental data from vegatables using different multivariate techniques, in *Progress in Flavor Research 1984*, Adda, J., Ed., Elsevier, Amsterdam, 1985, 131.

34. **Cortes, H. J., Pfeiffer, C. D., Jewett, G. L., and Ritcher, B.** E., Direct introduction of aqueous eluents for on-line coupled liquid chromatography-capillary gas chromatography, *J. Microcolumn Separ.*, 1, 28, 1989.

35. **Grob, K.,** On-line coupled high performance liquid chromatography-gas chromatography, *TrAC*, 8, 162, 1989.

36. **Nitz, S., Kollmannsberger, H., and Drawert, F.,** Determination of sensorial active trace compounds by multi-dimensional gas chromatography combined with different enrichment techniques, *J. Chromatogr.*, 471, 173, 1989.

37. **Van Hesse, V. G. and Grant, D.,** Automatic multidimensional capillary gas chromatography — on-line pre-separation and sample pretreatment, *Am. Lab. (Fairfield)*, 20, 26, 1988.

38. **Martens, M. and Martens, H.,** Near infrared reflectance determination of sensory quality of peas, *Appl. Spectrosc.*, 40, 303, 1986.

39. **Osborne, B. G. and Fearn, T.,** Discriminant analysis of black tea by near infrared reflectance spectroscopy, *Food Chem.*, 29, 233, 1988.

40. **Hall, M. H., Robertson, A., and Scotter, C. N .G.,** Near-infrared reflectance prediction of quality, theaflavin content and moisture content of black tea, *Food Chem.*, 27, 61, 1988.

GENERAL REFERENCES

1. **Smith, R. M.,** *Gas and Liquid Chromatography in Analytical Chemistry*, John Wiley & Sons, Chichester, 1988.

2. **Guiochen, G. and Guillemin, G. L.,** *Quantitative Gas Chromatography for Laboratory Analysis and On-Line Control*, Elsevier, Amsterdam, 1988.

3. **Westerlund, B. D., Ed.,** International symposium on coupled column separations: techniques for integrated multistage separation in liquid, gas and supercritical fluid chromatography and electrophoresis, *J. Chromatogr.*, 473, 313, 1989.

Chapter 4

PREPROCESSING: PREPARATION FOR DATA ANALYSIS

I. INTRODUCTION

Chemical data preprocessing includes all calculations needed to prepare the chemical data for analysis. Necessarily, this includes instrumental calibration, determination of sample extraction recoveries, the precision, and the accuracy for a given analytical procedure. The data needed to perform these calculations are acquired as a part of the procedure development and validation.

For most instrumentataion, the raw data do not represent meaningful quantities in relation to sample composition. A calibration function must be applied to transform the raw data into meaningful units. The overall calibration function consists of both the instrumental calibration and appropriate extraction recoveries. Reporting the results in absolute units such as parts per million (ppm) or weight percent is the best for comparison of samples.

In addition to calibration of responses, other calculations are used to condition the data mathematically. Data conditioning is necessary so that the data reduction routines are not artificially biased by differences in scale resulting from different measurement techniques, reporting units, or large magnitude differences for various compounds in the same sample (e.g., 1 ppm vs. 100 ppm). Data conditioning is also needed to make chemical and sensory data compatible for direct comparison. Finally, there are several ways of combining sensory and chemical data that help to add perspective for interpretation of the data reduction results.

II. ANALYTICAL PROCEDURE CALCULATIONS

A. INSTRUMENT CALIBRATION

This step is most often accomplished in the instrumental data system. The instrumental data system is the ideal place to do such calculations since it is programmed for that purpose. Good analytical technique requires that the procedure is calibrated each day that analyses are performed. For instrumental analyses, calibration can be either by the external standard or the internal standard procedure, whichever is appropriate. For wet chemical analyses, calibration is performed by reagent standardization. For physical measurements such as texture, a standard sample needs to be checked to ensure that no drift has occurred in the electromechanical transducers. The following sections outline general procedures for instrument calibration. Most of the discussions are focused on chromatographic applications. Calibration of nonchromatographic instrumentation uses many of the same procedures. For discussion purposes, peak area will be used as the measure of chromatographic peak magnitude. It is understood that chromatographic peak height or other data can be handled in an analogous manner.

The best way to report and analyze chemical data is in absolute units. This makes comparison of different samples straightforward. It also allows one to use the chemical data to guide product development/formulation efforts. The development of psychophysical functions, as described in Chapter 6, requires the knowledge of the concentration as well. Ideally one would like to develop psychophysical functions for all sensorially active compounds to provide the most detailed description of a sample. Practically, this is impossible, so the *patterns* of compound composition are interrelated by data analysis techniques.

Area percent is the default report format for most chromatographic data systems. Area percent expresses the area for each peak as a fraction of the sum of the areas for all peaks. This type of calculation results in data closure.[1,2] The high performance liquid chromatography (HPLC) data in Table 1 demonstrate the relationship between raw data, calibrated data, and

TABLE 1
High Performance Liquid Chromatographic Data Reported in Different Formats

Peak	HPLC peak areas				Calibrated result (ppm)				Area % calculation			
	1	2	3	4	1	2	3	4	1	2	3	4
1	96464	187080	70907	137862	6.35	12.3	4.66	9.06	5.98	8.62	8.18	8.46
2	30166	91817	34513	63775	2.20	6.70	2.52	4.65	1.87	4.23	3.98	3.91
3	15378	15780	2290	5938	1.07	1.10	0.16	0.43	0.95	0.72	0.26	0.36
4	12288	22613	6952	12284	4.56	8.39	2.58	4.56	0.76	1.04	0.80	0.75
5	742693	992949	483784	881183	47.5	63.5	31.0	56.4	46.1	45.8	55.8	54.1
6	11951	22776	2392	12787	3.09	5.90	0.62	3.31	0.74	1.05	0.28	0.78
7	84032	137388	48266	109531	32.7	53.4	18.8	42.6	5.22	6.33	5.57	6.72
8	252539	152705	48035	92345	94.2	57.0	17.9	34.4	15.7	7.04	5.54	5.67
9	109612	87628	24712	57156	72.3	57.8	16.3	37.7	6.80	4.04	2.85	3.51
10	11109	5321	5262	6782	19.0	9.10	8.99	11.6	0.69	0.25	0.61	0.42
11	242600	451452	128623	247694	381.	709.	202.	389.	15.1	20.8	14.8	15.2
12	2437	2800	1089	2074	0.47	0.54	0.21	0.40	0.15	0.13	0.13	0.12

the area percent format for reporting data. It shows the potential distortion in the relative level of an analyte that can occur when data are incorrectly reported. Different conclusions would be drawn about the similarities and differences between the samples depending on the data that is analyzed.

Inspection of Table 1 shows that there is a direct correspondence between the raw and calibrated data. The calibrated data were obtained by applying a calibration function for each peak. For this data, single point calibration was used to convert the peak areas to calibrated results. Peaks of the same magnitude reflect the same level of a given compound in different samples. The relative magnitudes of peaks within a sample vary for data in these two formats depending on the calibration factor for each peak.

For the area percent data, the calculated magnitudes of the individual compounds are in proportion to their peak areas *within the same* chromatogram. The relative levels of a compound in *different* samples may not be accurately described. Large peaks exert a strong influence on the calculation of area percent results. Comparing the relative magnitudes of peak 5 across the samples shows that sample 3, which has the lowest absolute level, would appear to have the most of peak 5 by the area percent method. For two samples that are identical, the area percent method yields the same results. However, for samples differing in composition, the area percent calculation might show that they are different, but the differences do not necessarily reflect the true relationships between the samples. Comparison of samples 3 and 4 shows that the true relationship in relative magnitudes is lost. By the area percent procedure, these two samples are virtually identical. The true relationship between the concentrations is shown to be about twofold in the calibrated data. Other inconsistencies are also present in this table. The interested reader should study the data to identify them. The area percent method or any calculation that results in data closure can potentially cause a large distortion and should be avoided when reporting data for chemometric studies.

When absolute calibration is not possible, meaningful comparisons of the relative levels of analytes in different samples can still be obtained. For such comparisons, a common basis for estimation of the amount of an analyte in each sample is needed. Means of obtaining both absolute and relative calibrations are discussed in the following sections.

1. External Standard Calibration

External standard calibration is appropriate when detector drift is not a problem, pure samples of the analytes are available for calibration of the instrument, sample extraction/preparation is very reproducible, and variations in injection volumes are minimized. It can be tedious when acquiring data for many analytes simultaneously, since a calibration curve for each analyte may need to be generated. External standard calibration is most commonly used in HPLC.

Calibration curves are usually linear of the form given in Equation 1:

$$\text{Area} = \text{slope} \times \text{concentration} + \text{intercept} \tag{1}$$

This relationship is developed by comparison of the peak areas (detector response) for standard solutions to the known concentration of the solutions. The standard concentrations must cover the range of possible concentrations in samples. The standards should be analyzed in random order along with the samples to check for instrumental drift during the course of sample analyses. Some instrumental data systems require that all standards be analyzed prior to the samples so the calibration curve can be calculated and used for calculation of amounts as each sample is analyzed. In such a situation, additional standards should be randomly placed in the analysis sequence to check for instrumental drift.

Least squares regression is commonly used to calculate the slope and intercept so that the concentration of the analyte in a sample can be calculated. The slope of the calibration curve is a measure of the sensitivity of the detector to the analytes and the intercept is an indication of the bias in the procedure.[3] Several goodness of fit criteria, to check the linearity of the calibration, are available as a check to ensure that the instrumental system is performing properly on any given day.[3,4]

Other external standard calibration procedures are sometimes included in instrumental data systems. Single point and multilevel calibration schemes are common. They do not develop a calibration curve; rather, they use the response to one or more standards to calculate a response factor (R_f) that typically has units of area counts per mass. For one point calibration, this response factor is multiplied by the peak area to yield the amount of analyte in a given sample. For multilevel calibrations, R_f for multiple standards containing different amounts of analyte are calculated, and the amounts of analytes in samples are determined by multiplying the peak areas by the R_f for the standard closest in peak area to that found in the sample. Such calibration schemes are acceptable as long as the researcher determines that they are yielding good results. This can be done by the random reanalysis of standards to check for system performance changes. Single point or multilevel calibration is not as rigorous as the least squares approach, but is easier to implement in data systems with limited computing power.

Calculation of the level of an analyte in a sample must take into account the amount determined from the analysis of the prepared sample (e.g., sample extract), and the amount of the original sample. Sample preparation can include extraction, dilution, or nothing. For extracted or diluted samples, a multiplication factor is needed as described in Equation 2:

$$\text{Amount} = C_p \times P_f/\text{sample weight} \tag{2}$$

where C_p is determined from the calibration data and P_f is the sample preparation factor. P_f is equal to the dilution factor or 1/extraction efficiency times a concentration factor for sample extraction procedures. Strategies for determination of P_f are discussed in Section II.B.

The external standard calibration scheme normally requires that a pure standard material is available for each analyte of interest. In studies of complex samples, it may not be possible to obtain a pure sample of each analyte or even to identify each unknown. This situation should not prevent the chemist from gathering information on all possible compounds, as the potential exists that the data analysis will suggest the importance of previously unknown compounds. Under these circumstances, a couple of different means of getting good comparative data can be used. These approaches do not allow the determination of absolute amounts, but will provide data that can be compared to identify similarities and differences between samples. The simplest approach is to just use peak areas as long as it can be shown that the analysis procedure satisfies the reproducibility requirements mentioned above. To *estimate* absolute amounts, one can determine the chemical characteristics of unknowns with respect to known compounds and use calibration data for the known compounds to estimate the levels of unknowns. An example will demonstrate this concept.

Natural beverages contain series of related compounds that vary in the number and position of substituents in relatively large polymeric molecules. Flavonoids and polyphenols are examples of such compound series. Within these general classes, several subclasses of compounds exist. Compounds in each subclass have a similar chemical backbone. This basic chemical structure gives all those compounds similar chemical characteristics. Choosing a detector that measures the presence of that backbone structure allows one to estimate the level of unknown compounds.

Lunte describes a detailed compound classification scheme based on this approach.[5] For the use of a UV-VIS detector, related compounds will contain the same chromophore, will exhibit the same absorption maxima, and have similar extinction coefficients.[6] Use of a UV-VIS diode array detector allows one to determine the spectral characteristics of unknown compounds and functionally classify them based on the spectral characteristics of known compounds. The use of a second detector, in this case an electrochemical detector, provides an even more detailed classification scheme. From such experimental data, one can obtain a functional classification of unknowns. Calibration data for known related compounds can then be used to obtain a reasonable estimate of unknown levels. Determination of the specific identity of unknowns requires additional spectroscopic work.

This approach does not work with all detectors. The refractive index detector cannot be used for compound classification, but may be useful after such classification has been obtained. The user would have to show that the RI response for related compounds is the same. The fluorescence detector can be used for classification of unknowns by scanning both the excitation and emission spectra. Functional classification is obtained by comparison of these spectra with that for known compounds. Differences in quantum efficiencies can limit the accuracy of results.[7]

2. Internal Standard Calibration

Internal standard calibration is used when there is a good chance of variation being introduced in an analysis procedure. An internal standard can compensate for variations in extraction efficiency, injection volume, and detector drift. It cannot compensate for sampling errors, sloppy technique, and changes in relative detector response. The internal standard should be stable, behave in a predicitable manner during extraction and analysis, and be susceptible to the same procedure variations as analytes. It must not interfere with the signals (e.g., coelute) for the analytes. Internal standard calibration allows calculation of levels for many analytes when the response of each analyte is calibrated against that for the internal standard. For the analysis of complex samples with many analytes, it is the method of choice for both HPLC and GC.

The general equation for calculation of amounts based on an internal standard is given in Equation 3:

$$\text{Amount}_a = \frac{\text{Area}_a}{\text{Area}_{is}} \times R_{f,a} \times \frac{\text{Amount}_{is}}{\text{Sample wt.}} \times P_{f,a} \qquad (3)$$

The amount of an analyte, a, in the original sample is determined from the ratio of the measured response (Area) for both the analyte and the internal standard, the relative detector response factor, $R_{f,a}$, the amount of the internal standard added to the sample, the sample weight, and any preparation (e.g., dilution, concentration, and extraction recovery), $P_{f,a}$. In

Equation 3, the measured peak areas and sample weight as well as the amount of internal standard are quantities defined at the time of sample analysis. $R_{f,a}$ and $P_{f,a}$ must be determined during method development. The values used for $R_{f,a}$ and $P_{f,a}$ will affect the calculated results as discussed in the following sections.

The simplest internal standard calculation is to calculate the ratio of the peak area for each analyte to the peak area of the internal standard. This calculation compensates for variations described above but does not allow for the determination of absolute amounts. If one assumes that the relative response factor, $R_{f,a}$, for the analyte and internal standard is 1.0, the calculation yields an estimate of the analyte level.

Unknowns are usually assigned an R_f of 1.0. For the FID, this is a reasonable approximation.[8] The FID response factor falls in the range of 0.8 to 1.2 for most compounds; for hydrocarbons, it ranges from 0.96 to 1.12. Assuming an FID response factor of 1.0 for most compounds will allow the calculation of amounts that are in the right ballpark. The major limitation for the FID exists for compounds with a significant number of oxygen atoms (e.g., acids, esters, aldehydes, and ketones) that tend to quench the signal. Compounds with a low carbon-to-oxygen ratio, such as methanol and formic acid, have FID response factors of 0.23 and 0.01, respectively. Carbon dioxide does not show an FID response at all. Most of the other GC detectors are more compound specific. For these detectors, $R_f = 1.0$ is used as a matter of convenience and may not necessarily be a good estimate. Significant amounts of work may be required before generalizations about detector response can be made.

The concept of functional classification of unknowns, as discussed for external standard calibration, is also applicable for internal standard calibration. If the functional classification of the unknown compound is possible, then an estimated $R_{f,a}$, based on that for similar known compounds, can be used to yield an estimate of the unknown level.

The most rigorous way is to determine $R_{f,a}$ for each compound. $R_{f,a}$ should be determined under actual analysis conditions. For example, the response may vary among split, splitless, and on-column injection if there is discrimination in the injector. Determination of $R_{f,a}$ for a compound requires a standard of known purity for each compound. Standard solutions, of known concentrations, are analyzed to calculate r_f (area/mass) for both the analyte and the internal standard. $R_{f,a}$, the relative response factor, is then calculated as the ratio of the r_f for the internal standard and the analyte as given in Equation 4:

$$R_{f,a} = \frac{r_{f,is}}{r_{f,a}} = \frac{area_{is}}{mass_{is}} \times \frac{mass_a}{area_a} \tag{4}$$

The discussion so far has centered around the relationship of the levels of the same analyte in different samples. How do the same considerations affect the representation of the relative levels of different compounds in the same sample? For the most rigorous calculation, where $R_{f,a}$ has been determined for each analyte, the calculated results will reflect the true relationships between two analytes. This is the ideal situation as it is the most accurate description of sample composition. In the absence of this information, the true relationship between two analytes will not be known.

When $R_{f,a}$ has not been determined, a limitation exists in the data. For the FID, this limitation is not severe and represents variations around the noise level of typical analysis procedures. For other detectors, it can be much more severe, depending on how different $R_{f,a}$ is from 1.0. In the situation where the response factors cannot be determined for each analyte, all is not lost since the *fingerprint*, or *pattern*, for a given sample will still be unique and define the composition that elicits a given sensory response. Similar patterns should represent similar sensory responses, and changes in patterns should relate to changes in sensory response. If they do not, there are two probable reasons: either the compounds that changed are not important to the sensory impression and/or the compounds that define the sensory impression are not being measured by the analysis procedure.

B. DETERMINATION OF EXTRACTION RECOVERIES

Instrumental calibration permits the determination of the levels of analytes in the solution that are introduced into the instrument. For the filter, dilute, and shoot (FDS) sample preparation procedure, no extraction losses are encountered. The amount in the sample is calculated from the amount in the analysis solution by multiplication with the dilution factor. This is straightforward. For procedures where the sample is extracted prior to analysis, knowledge of the extraction recoveries or extraction efficiencies is needed in order to calculate absolute analyte levels. The results of the analysis of the final extract must be multiplied by a number that compensates for any losses during the extraction and extract concentration process.

The scheme for the determination of extraction recoveries depends on the technique used. Several different general strategies exist. They include the methods of *standard addition,* comparison of *added vs. found* amounts and *exhaustive* sample extraction. The conditions for application of each technique varies.

The method of standard addition can be used when it is impossible to acquire a sample matrix that is free of the analytes of interest. Although normally used for the determination of analyte amounts, standard addition can also be used to determine recoveries of analytes for the sample extraction/preparation procedure. Aliquots of the same sample are used for multiple determinations. First, the sample is analyzed as is to determine the amounts of analytes present. Known amounts of analytes are then added to a fresh sample aliquot and analyzed to determine the original plus added amounts. The addition of at least two different known amounts is used to generate a standard addition curve as shown in Figure 1. The amount of analyte(s) added is determined by direct analysis of an aliquot of the spiking solution. The slope of the line is the recovery of the analyte, including both the extraction efficiency and any losses for the extract concentration step.

A rigorous derivation of the generalized standard addition method is contained in References 9 and 10. The standard addition procedure has the advantage that the analysis is carried out on exactly the same matrix as the original sample. Care must be taken to ensure that dilution of the sample by the added standard is minimized or compensated for by an internal standard. When the spiking solution contains multiple analytes, data for each analyte can be obtained with a minimum number of sample analyses.

The direct comparison of added vs. found amounts is a straightforward approach to determine recoveries if a blank sample matrix is available for spiking experiments. It is a special case of the standard addition procedure since analysis of the initial sample, in this case the blank matrix, results in none found. The ratio of the amount found to the amount added (slope of the added vs. found curve) is the recovery of the analyte. For formulated samples, this procedure can be easily implemented, as it is possible to obtain samples of a blank matrix. For natural products, this procedure is more difficult as the flavor compounds must be stripped from an authentic sample prior to addition of the standard. It is important that the matrix used for the recovery experiments is as close as possible to the authentic matrix to avoid matrix effects.

The complete removal of analytes from a sample by successive extractions is readily applicable to method development with SPE cartridges. A schematic diagram is given in Figure 2. The sample is extracted as if being analyzed normally in the first step. The supernatant (sample matrix plus unretained analytes) from the sample is then passed through a second SPE cartridge to recover the unretained analytes. Successive extractions can be repeated until no more analytes are found. The fraction of the analytes that are removed by each cartridge can be calculated. The third part of the scheme determines the retention of the analytes for each SPE cartridge. Each cartridge is eluted with additional aliquots of solvent to

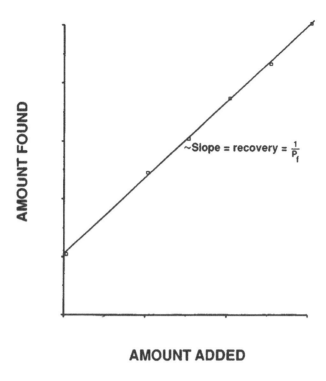

FIGURE 1. Determination of recoveries by the standard addition procedure.

determine if analytes are left on the cartridges after normal solvent elution. The recovery of the analytes in each solvent aliquot can then be determined.

The three-step sequence makes it possible to track down the location of all fractions of the analytes and to optimize the extraction conditions by determination of the required SPE capacity and the appropriate solvent aliquot size needed for analyte elution. Practical limitations may prevent using the conditions that provide 100% recovery for all analytes. In this circumstance, the data gathered allow for determination of the recovery for each analyte. This approach can be combined with standard addition for further validation of sample extraction recoveries. For other liquid-liquid or liquid-solid extraction techniques, a similar strategy can be used. The sample is extracted normally, followed by a second and a third extraction. The amount of material contained in each extract is then determined.

Validation experiments can be tedious and time consuming for samples containing many analytes. Typical problems include the acquisition of individual chemicals, checking their purity, calibration of the instrumentation for all the analytes, and the availability of a blank sample matrix for the spiking experiments. The standard addition procedure can be used to determine analyte recovery when a blank matrix is not available. Added-found studies are a straightforward way to determine analyte recoveries when blank sample matrices are available. Multiple, successive extractions are ideally suited for method validation with SPE cartridges. The above mentioned schemes are not difficult to perform when just a few analytes are to be determined. However, when many compounds must be quantitated simultaneously, they require sufficient planning and care during execution so that spiking solution(s) containing all the analytes can be prepared in the right concentration range.

C. DETERMINATION OF PRECISION AND ACCURACY

Precision is a measure of how reproducible an analysis procedure is. It is composed of the

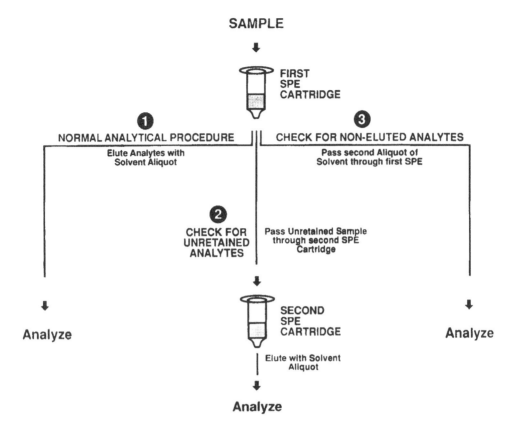

FIGURE 2. Scheme for determination of recoveries with SPE cartridges.

individual variabilities from the sampling, extraction, and instrumental or chemical analysis procedures. Usually, the overall precision (s_T), resulting from the combination of all steps is of the most interest. However, if the overall standard deviation is too large, it may be necessary to determine the effect of each step in the procedure to track down the cause of the large variation.

Precision is estimated by calculation of the standard deviation, s, for multiple test results. The standard deviation of a sample population is calculated according to Equation 5.

$$s_i = \left(\frac{\Sigma(X_{ij} - \overline{X}_i)^2}{(n-1)} \right)^{1/2} \tag{5}$$

where s_i is the sample population standard deviation for analyte i, X_{ij} is the value for analyte i in sample j, and $\overline{X}_i$ is the average for analyte i across all n samples. It is a measure of the deviation of individual analysis results from the mean for all the analyses. Since the deviations are squared prior to summation, it is always positive. The standard deviation has the same units as the original variables. For ease of comparison, the absolute standard deviation is normalized to the mean value to yield the relative standard deviation (RSD) or coefficient of variation (CV) as given in Equation 6.

$$RSD_i = s_i/\overline{X}_i \times 100 \tag{6}$$

The RSD is a dimensionless quantity and allows comparison of results that are reported in different units or scales.

The square of the standard deviation is defined as the sample *variance*. The overall variance for several discrete steps in a procedure is determined by the uncertainties in each step of the procedure. For sample analysis procedures, the total variance can be approximated by the sum of the relative variances for each step in the procedure:[3,10]

$$RSD^2_T = RSD^2_{samp} + RSD^2_{ext} + RSD^2_{inst} \qquad (7)$$

The overall standard deviation or variance is of utmost interest and is usually determined by analyses of multiple samples. If s_T is large, then each step in the procedure must be evaluated to determine which part needs to be brought back into control. The individual variances are determined by experiments designed to isolate each step from the effects of the others. This can be done piecewise by holding all the procedure steps constant except for the one that is being evaluated.

Instrumental precision (s_{inst}) is determined by replicate analyses of the same sample extract. This allows calculation of s_{inst} measuring variations caused by sample injection, detection, and measurement of detector response. Instrumental precision is the most complex of the three since it is composed of multiple steps, each of which can contribute variability. It is difficult to decompose into its component parts since the measurement of each is dependent on the others. Once the instrumental precision is known, the precision of the extraction procedure (s_{ext}) can be determined by replicate analyses (extractions and instrumental analyses) of multiple samples from the same container, assuming that the material in the container is homogeneous. s_{ext} can then be calculated since s_{inst} and s_T have been determined. In an analogous manner, the variation caused by sample-to-sample differences (s_{samp}) can be determined by analyses of multiple samples (e.g., containers) followed by calculation according to Equation 7.

To summarize, calculation of the relative standard deviation (RSD), also called the coefficient of variation (CV), yields an estimate of the precision of the analysis procedure. Most procedures should be able to operate with an overall RSD of 10 to 20%. If the RSD is greater than 20%, each step in the procedure needs to be evaluated in order to identify and obtain better control over those that are causing the large variation.

Accuracy refers to how close the reported value is to the true value. The accuracy of a calculated result depends on the precision of the procedure, interferences in the determination, effects of the matrix, instrumental calibration, and losses that occur during sample preparation. Proper method validation yields the necessary information to provide accurate results.

Different schemes exist for determination of the accuracy for a given procedure.[10,11] Accuracy is usually measured by *bias* in the found vs. known results. Bias in reported values can be either fixed or proportional. *Fixed* bias means that all results deviate from the true value by a constant amount. An example of this is a nonzero intercept in a calibration curve caused by interference. *Proportional* bias means that the results deviate from the true value by a percentage. As the true value changes, so does the absolute deviation in the reported value. Larger values have larger absolute errors. Once any bias has been identified, a correction can be made to yield an absolute amount present in the sample.

D. INSTRUMENTAL PERFORMANCE TESTS

It is advisable for the scientist to check the performance of the instrumental system on a regular basis. Such tests, called *system suitability* tests, confirm that everything is operating properly and help to assure the quality of the data obtained on a given day. Different checks can be performed on data that are gathered during normal sample analyses. The analysis of standard samples can be used to determine the retention time reproducibility, calculate the chromatographic column efficiency (theoretical plates/meter), and check the linearity of calibration curves or the consistency of detector response factors. Each test helps to monitor different parts of the instrumental system.

Retention time reproducibility is of primary importance when analyzing complex samples with many analytes. It is a necessity, as the retention time is the main criterion used for identification of analytes and the alignment of chromatographic peaks for further data analyses. Significant variations in retention times may be indicative of changes in instrumental operating conditions. For HPLC, retention time shifts may be caused by use of the wrong mobile phase, changes in mobile phase flow rates (pump problems or leaks), or inadequate column reequilibration after gradient elution. For GC, such deviations are indicative of changes in gas flow rates (incorrect operating pressures, septa or other leaks, or partial blockage of flow paths) or incorrect temperature programs. For both techniques, significant drift in retention times may also be an indication of column deterioration.

Calculation of the column efficiency, as in Equation 8, provides a direct indication of changes in column performance.

$$N = 16(t/W)^2 \qquad\qquad (8)$$

where N is the number of theoretical plates in the column, t is the retention time, and W is the width at the base of the peak. W is measured by extrapolation to the base line from the relatively straight sides of the peak and expressed in the same units of time as t. Most new columns come with test results indicating the test mixture used, analysis conditions, and the method of efficiency calculation. It may be useful to determine the efficiency of a new column under actual analysis conditions to see how it compares to the manufacturer's test and for future reference.

Calibration curve linearity and detector response factor checks help to identify problems with detector performance and injection technique (for GC). Periodic reanalysis of a stable check sample is another way to monitor overall system performance. Any of the problems mentioned above can result in significant changes in reanalysis results. In addition, this last test also checks the performance of the instrumental data system to ensure that the same conditions are being used for peak integration and calculations.

The limits of acceptability for the system suitability tests are usually developed during method validation and should be realistic. Too tight limits may cause acceptable data to be rejected, while too loose limits may allow the use of data that contains artificial variations that do not reflect true differences between samples. Successful completion of these instrumental performance tests provide confidence that equivalent data are being gathered with each set of samples.

III. DATA HANDLING CONSIDERATIONS

Modern analytical laboratories can routinely generate large volumes of detailed data for complex samples. To obtain the most information from this data, systematic organization and reduction is needed. Data handling consists of data collection, organization, and creation of data bases. These tasks are best accomplished with the use of automated instrumentation and data handling procedures. Automation offers the advantages of speed and accuracy and greatly lowers the energy barrier for routine application of chemometrics. Data collection, including sample extraction and analysis, normally takes longer than the actual data analysis. Manual organization and creation of databases greatly increases the amount of time and effort required to prepare for data analysis.

Figures 3 and 4 graphically demonstrate two different approaches to data handling. Either procedure can implement the same three steps. Both assume the existence of computerized data acquisition as a part of the instrumental system.

The typical arrangement in Figure 3 consists of a stand alone instrumental data acquisition system that requires a data transfer interface for serially uploading the data for each sample.

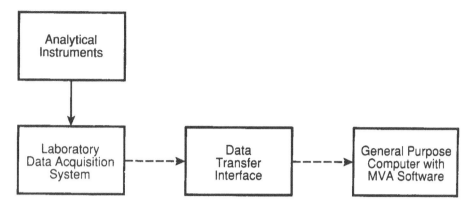

FIGURE 3. Typical data acquisition and analysis configuration.

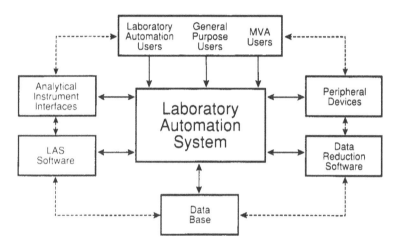

FIGURE 4. Ideal data acquisition and analysis environment.

This arrangement is relatively easy to implement but requires much user interaction to move data from one system to another and to organize the database for analysis. Many times, the data transfer interface is the chemist or laboratory technician who manually organizes, tabulates, and enters the data into the computer where the data analysis will occur. A better arrangement is when the data transfer interface is electronic and information is uploaded via RS 232 or other standard protocol to a host computer.[12] Personal computers (PCs) and PC-based instrumental data systems are useful for this step since commericial database managers make it easy to manipulate large quantities of data.[13] It is even possible to use the PC for the final data analysis, as multivariate software packages are now available for PCs.

Figure 4 describes a completely integrated system where all functions are performed in the same computer. Such an approach offers all the advantages mentioned above in addition to supporting multiple instruments and users simultaneously. Multiuser laboratory automation systems (LASs) can be configured for such operation. The vendor software provides for the data acquisition and possibly the data management (LIMS). However, at this time we are not aware of a commercial LAS system that offers the data analysis features as a part of the integrated package. (These capabilites are available in a few commercial spectrometer data systems). General-purpose statistical software packages are available for various laboratory computer systems. Data analysis packages have been developed privately and integrated into laboratory computer systems.[14] The major disadvantage of this approach is the need for the custom development and/or installation of the data reduction software.

For most users the data handling procedures will fall somewhere between the two described in Figures 3 and 4. Final configurations and manipulation schemes will be determined by the availability of money, hardware, and software. Many times, the manual approach can be used to demonstrate the benefit of chemometrics as a justification for investment of future time and money. With the common availability of PCs, users should seriously consider their use and take the time necessary to evaluate all their capabilities. Commercially available PC software packages will probably offer capabilities approaching total integration in the near future.

A. DATA COLLECTION

Computer-based data collection sytems are an integral part of most modern instrumentation. No matter what instrument is being used to characterize a sample, computerized control and data acquisition is a necessity for chemometric studies.

For chromatographic instruments, electronic signal integrators are usually used and are greatly preferred to strip chart recorders. Stand alone, general purpose integrators, and those supplied as an integral part of a system generally operate on the same principles and can provide equivalent data. Quantitation of peak magnitudes can be done by the use of peak areas or peak heights. Although peak areas are used most often, peak heights can prove advantageous in certain circumstances. Peak areas are less influenced by changing instrumental and chromatographic parameters, less sensitive to drifting baseline, more sensitive to asymetric peaks, and most appropriate for mass sensitive detectors (e.g., FID, PID, and MSD). Peak heights are simple to measure, less sensitive to loss of chromatographic resolution, less sensitive to tailing peaks, and most appropriate for concentration sensitive detectors (e.g., IRD, RAD, and TCD).

Many data systems allow the user to perform the first step of data preprocessing in the instrumental data system. The first step is transformation of the data into meaningful units. This is usually accomplished by calibration of the data system to be able to report the data in absolute or relative (to an internal standard) units. The data system should be used to its fullest extent, since it is designed to perform chemical calculations. Most general-purpose data analysis software packages do not contain modules that provide for such calculations. Even if the available data acquisition system does not have electronic data transfer capabilities, automatic completion of this important step in the process greatly expedites the overall process.

B. DATA ORGANIZATION

In order for the chemometric software to do its job, proper organization of the database is absolutely necessary. The data must be properly formatted in a file so the software can access it. Data for chemometric studies is represented as two-dimensional *arrays*. Each sample in the *matrix* is described by a *vector* of data points. Individual variables (e.g., peak magnitudes) are the *elements* of a data vector. A one-to-one correspondence must exist for each variable for each sample; that is, a given variable must represent the same measurement for all samples. If it does not, any trends found in the data will just be noise. For chromatographic data, this means that all peaks must be properly aligned. Peak alignment can be a major task for capillary GC profiles containing 100 or more peaks. A database containing only 20 samples can have 2000 or more data points. Data point alignment is frequently the most demanding of the tasks to be performed; it can be accomplished either automatically or manually. Usually, a combination of the two is required.

Instrumental data systems can provide some of the peak alignments. The simplest way is to program the system to only report "identified" peaks. This approach yields less data and frees the user from manual peak matching. However, it also eliminates some potentially interesting data. Alternately, for systems with higher levels of programmability (e.g., BASIC), an algorithm can be developed to yield a matched output. General peak matching algorithms are not commercially available and must be custom developed.[12-14]

A few comments on chromatographic peak matching are in order. Retention times are used as the criteria for peak matching. Therefore, retention time reproducibility (as discussed in instrumental performance tests) is as important as good analytical precision. Normal variations in retention times are used to establish *windows* for peak matching. When retention times vary beyond the normal windows, retention time indexing schemes can be used to adjust for the shifts.

The approach that we have found to be useful is to create a template that contains all the possible peaks found in all the samples. Matching is accomplished by comparison of a given sample to the template. A match is found when a peak is present that falls within the expected retention time window. A good starting point is to use the sample with the most peaks as the initial template. In the process of matching peaks, the template will grow as more samples are added. The probability is high that a given sample will not contain all peaks. When a peak is not present in a sample, the value must be set to zero to hold the place for that variable in the sample vector. Since the data analysis routines are mathematical, any nonnumerical character is not acceptable as a place holder.

Other points of interest when organizing a database include how data from different measurements (HPLC, GC) will be combined. Depending on software and analysis schemes, it may be best to create a master file containing all the data and then to work with subsets as needed. Alternately, several smaller files can be developed and ultimately combined as needed. Planning must also include how sensory data will be handled.

To summarize, the user must be familiar with data handling characteristics of the data reduction software. Most data reduction packages offer some capabilities for data editing and manipulation. Time is well spent in advance to determine the requirements and capabilities of selected software packages. The combination of capabilities from several different packages may be needed to perform all the desired functions. Available PC software should be considered before custom development is started.

IV. DATA TRANSFORMATIONS

Both univariate and multivariate data reduction techniques are designed and implemented to identify similarities and differences in the magnitudes of variables. Univariate techniques are based on mathematical hypotheses that two sample populations are the same or different based on the distribution of the sample populations. Tests, such as the *t-test*, are performed to determine if significant differences in the sample distributions exist; that is, individual variables are tested to see if they can discriminate between samples.

Multivariate techniques have been referred to as soft models. They are not designed to determine, within certain confidence limits, if two samples are the same or different; rather, they are designed to identify the presence or absence of simultaneous trends or patterns in multiple variables across samples. Because multiple variables are compared simultaneously, one must be sure that artifactual differences in the variables are minimized. Data conditioning eliminates these artifactual differences so that true similarities and differences can be identified.

A. DATA CONDITIONING

For both univariate and multivariate analyses, data conditioning prevents bias caused by differences in the absolute magnitudes of variables. These artifactual differences can arise from data obtained on different instruments or by different analysis (GC vs. HPLC) techniques and when comparing sensory to instrumental data. This phenomenon, caused by different measurement scales and reporting units, can result in large magnitude variables receiving more weight in certain calculations when, in fact, the relative variations in the larger numbers may not be as great as those in smaller numbers.

The most common form of data conditioning is to calculate *standardized deviates*. Other synonomous terms are *standard scores, Z-values,* and *autoscaled* variables, the latter term being found primarily in the chemical literature. This calculation, which results in dimensionless units, removes the differences caused by different measurement scales or reporting units and allows all variables to be compared on the same standard normal distribution.

Standard scores are calculated for each variable (e.g., chromatographic peak) in each sample. They are obtained by subtracting the mean for a variable (across all samples) from the value for the sample and dividing the difference by the standard deviation of the variable as given in Equation 9:

$$Z_{ij} = (X_{ij} - \bar{X}_i)/s_i \tag{9}$$

where X_{ij} is the value of the ith variable for the jth sample, $\bar{X}_i$ is the mean for variable i across all samples, and s_i is the standard deviation for variable i. This calculation transforms the original data into a standard normal distribution. The standard normal distribution has a mean of 0, a standard deviation of 1, and a range of approximately -3 to $+3$. The range results from the fact that 99.7% of the data in a normally distributed population falls within three standard deviation units of the mean.

This calculation maintains the relative magnitudes of the variables *between* samples, but changes the relative magnitudes of variables within a sample. For example, let variable X_1 have a true relationship with variable X_2 such that $X_1 = 10X_2$. Z_1 will have a relationship with Z_2 of $Z_1 = Z_2$. The tenfold difference will be removed. If the tenfold difference was caused by differences in reporting units or measurement scales, this is good since Z_1 will not be weighted 10 times more in magnitude calculations. However, if the difference is real and significant, then information is lost. It is still appropriate to use the Z scores for the data analysis. However, during the interpretation, the researcher must consider the significance of any magnitude differences in the original data.

An example may help clarify this point. Suppose that compounds X_1 and X_2 have the same sensory threshold. Let $X_1 = 10X_2$; from above, $Z_1 = Z_2$. Both compounds would show the same trend in the data analysis. However, since $X_1 = 10X_2$, X_1 is more likely to be sensorially significant than X_2. The researcher must include this information when interpreting the data analysis results. This could be considered to be part of the verification phase in the data analysis.

The previous discussions about the use of multiplication (e.g., calibration) factors, when reporting data, are put into perspective when considering the calculation of Z values; that is, the effect of multiplication factors (e.g., R_f) that change the value of a given variable, say X_1, by the same amount in all samples is canceled by the calculation of Z values. This is why ratios to an internal standard as well as absolute amounts can be used for sample comparisons. The data analysis results obtained for ratios to an internal standard or absolute amounts will be the same. Absolute amounts do have importance when considering relative levels compared to sensory thresholds and for guiding product formulation work.

B. THRESHOLD SCALING

Data analysis techniques are designed to find differences in samples. The contribution of compounds, where differences do not exist, may be overlooked if only the results of the data analysis are used. Threshold scaling provides perspective as to whether or not a given compound is of potential sensory significance. It helps to ensure that potentially important compounds are not overlooked.

Threshold scaling is a straightforward calculation as given in Equation 10:

$$C_{Ta} = C_a/T_a \tag{10}$$

where the concentration of a compound (C_u) is just divided by the absolute threshold (T_a) level of the compound. This calculation can only be performed after determination of absolute compound levels.

Odor and taste threshold data are available in the literature.[15] Division by the sensory threshold does not change the relative magnitudes of a compound in different samples. It does change the relative magnitudes of compounds within the same sample. Compounds that are many orders of magnitude above their threshold are likely to have an individual impact as well as a combination effect with other compounds. Compounds below their threshold will not have much impact by themselves. However, in combination with other compounds, they could contribute to the overall sensory sensation. Alkyl pyrazines are a good example of this phenomena.

Threshold scaling is analogous to the GC/sniffport technique called CHARM analysis.[16] It provides a magnitude estimation of the potential intensity of a compound, but does not guarantee the importance of that compound. Conversely, the absence of a sensory response $(C_{T_a} < 1)$ or no odor detected at the sniffport does not mean that compound does not contribute to the overall sensory impression. Both are qualitative tools that can be used to add to the understanding and interpretation of the data gathered for a series of samples.

Several reasons exist why these two approaches are not quantitative and absolute. Threshold scaling just tells if the level of a compound is above the sensory threshold. The real impact of a given level of a compound can only be determined by experimentation. Levels need to be checked by spiking experiments or the psychophysical function needs to be determined. Compounds with relatively flat psychophysical functions need much larger changes than compounds with steep functions to elicit a sensory change. Additionally, the synergistic effect for mixtures of compounds needs to be considered in complex samples.

The results of a GC/sniffport analysis depend on many variables that must be considered when interpreting the data. The ability to detect a compound at the sniffport depends on the sensory aroma threshold in the GC effluent, the amount of the compound in the sample extract, the amount injected on the column, the amount eluting from the column, and the transfer efficiency of the external sniffport apparatus. Calculations need to be done to put these amounts into perspective in relation to the amounts in the sample. Simply calculating relative magnitudes based on the sniffport response can be misleading if other parameters are not taken into account.

V. SUMMARY

The researcher needs to be aware of the information in his data. Proper procedure validation and calibration is a necessity for good quantitative work. Raw data must be appropriately transformed prior to the data analysis. The most information is obtained from the knowledge of absolute compound levels. Physical organization and creation of the database is a major step in the data analysis process. Database creation should be as automated as possible. Conditioning of the data prior to an analysis is required to prevent artificial differences from influencing the results. Additional information, such as sensory thresholds help in the interpretation of the data analysis results.

REFERENCES

1. **Johansson, E., Wold, S., and Sjodin, K.,** Minimizing the effects of closure on analytical data, *Anal. Chem.*, 56, 1685, 1984.
2. **Wold, S., Abano, C., Dunn, W. J., III, Edlund, U., Esbensen, K., Geladi, P., Hellberg, S., Johansson, E., Lindberg, W., and Sjorstrom, M.,** Multivariate data analysis in chemistry, in *Chemometrics — Mathematics and Statistics in Chemistry,* Kowalski, B. R., Ed., D. Reidel, Dordrecht, The Netherlands, 1984.
3. **Caulcutt, R. and Boddy, R.,** *Statisitcs for Analytical Chemists,* Chapman and Hall, New York, 1983.
4. **Burgard, D. R. and Kuznicki, J. T.,** Correlation and regression, in *Chemometrics: Chemical and Sensory Data,* CRC Press, Boca Raton, FL, 1990, chap. 9.
5. **Lunte, S. M.,** Structural classification of flavonoids in beverages by liquid chromatography with ultraviolet-visible and electrochemical detection, *J. Chromatogr.,* 384, 371, 1987.
6. **Ewing, G.W.,** *Instrumental Methods of Analysis,* 3rd ed., McGraw-Hill, New York, 1969, 50.
7. **Delcour, J. A., Vinkx, C. J. A., Vanhamel, S., and Block, G. G. A. G.,** Combined monitoring of UV absorbance and flourescence intensity as a diagnostic criterion in reversed-phase high performance chromatography separations of natural phenolic acids, *J. Chromatogr.,* 467, 149, 1989.
8. **Dietz, W. A.,** Response factors for gas chromatographic analysis, *J. Gas Chromatogr.,* 2, 68, 1967.
9. **Jochum, C., Jochum, P., and Kowalski, B. R.,** Error propagation and optimal performance in multicomponent analysis, *Anal. Chem.,* 53, 85, 1981.
10. **Massart, D. L., Vandeginst, B. G. M., Deming, S. N., Michotte, Y., and Kaufman, L.,** *Chemometrics: A Textbook,* Elsevier, Amsterdam, 1988, chap. 8.
11. **Mark, H., Norris, K., and Williams, P. C.,** Methods of determining the true accuracy of analytical methods, *Anal. Chem.,* 61, 399, 1989.
12. **Liardon, R. and Spadone, J. C.,** Coffee aroma investigation by combined capillary GC headspace analysis and multivariate statistics, *Colloq. Sci. Int. Cafe (C.R.),* 11th, 181, 1985.
13. **Holcombe, H. E. and Guichon, G.,** Symphony in the analytical laboratory, *Chemometrics Intelligent Lab. Syst.,* 1, 91, 1986.
14. **Burgard, D. R. and Anast, J. M.,** Characterization of flavor systems — automation and multivariate analysis, in Symp. on Chemometrics, ACS National Conference, Anaheim, CA, September, 1986.
15. **Fazzalari, F.A.,** *Compilation of Odor and Taste Threshold Values Data,* American Society for Testing and Materials, Philadelphia, 1978.
16. **Acree, T. E., Barnard, J., and Cunningham, D. G.,** A procedure for the sensory analysis of gas chromatographic effluents, *Food Chem.,* 14, 273, 1984.

GENERAL REFERENCES

1. **McDowell, R.D., Ed.,** *Laboratory Information Management Systems: Concepts, Integration, Implementation,* Sigma Press, Manhasset, NY, 1987.
2. **Head, M. and Smith, K.,** A network data acquisition system, *Int. Lab.,* 18, 40, 1988.
3. **Papas, A. N.,** Chromatographic data systems: a critical review, *Crit. Rev. Anal. Chem.,* 20, 359, 1989.

Chapter 5

THE CHEMICAL SENSES

I. INTRODUCTION

This chapter reviews the functional anatomy and perceptual processes of the chemical senses: taste, smell, and the common chemical sense. While it is possible to collect sensory data without this knowledge, interpretation of that data is not. In sensory evaluation studies, the human subject is often viewed as an instrument which is used to determine some property of the product samples being studied. Since the product is of primary interest, it is often easy to overlook the properties of the instrument. It is those properties, however, which determine the type and quality of data that is obtained. As an analogy, consider a liquid chromatograph. This instrument can be used to determine the concentrations of specific chemicals in a sample, the relative amounts of several chemicals in the sample, or even the total number of different chemicals present. The type and quality of data obtained depends on several variables. These include the way the samples are prepared for injection into the instrument, the solvent used to extract the samples and in which they are dissolved, the material in the column, the pressure at which the samples are pumped through the column, etc. A change in any one or combination of these variables will affect the data obtained in measurable ways. In a similar fashion, the human instrument possesses properties, which are not determined by the investigator but are brought to the situation, that affect the data obtained in a sensory evaluation study. These properties, not all of which are clearly understood, are determined by functional operating characteristics of the sensory systems. The analogy between a human sensory instrument and a liquid chromatograph cannot be carried too far, however, because unlike the passive chromatograph the human instrument is an active instrument. The information fed into the human instrument is actively processed such that the output of the instrument does not always, or even usually, stand in one-to-one relationship with the input. This processing begins at the earliest stages of the perceptual process, interaction of sample with sensory receptors, and continues through the last stages of the process in which the subject makes some sort of judgment about the sample (e.g., quality, intensity, or hedonic tone of the sample). The sensory properties of the human instrument are also affected by higher-order properties called cognitive processes which affect the data and, perhaps of greatest importance, the ability of the subject to provide the data of interest. To properly interpret output from a sensory system, some knowledge of that system is absolutely necessary. Fortunately, the study of sensory processes is a very active area and several excellent texts and handbooks are available.[1-8] The present discussion will focus on the chemical senses and specifically on the properties of those senses which have the most salient impact on the collection and interpretation of sensory evaluation data.

II. TASTE

A. ANATOMY OF THE TASTE SYSTEM

Taste is the sensation which is evoked by interaction of stimulus molecules with specialized receptor cells contained in taste buds. These taste buds are, in turn, contained in structures called papillae which are located primarily on the tongue. However, taste buds can also be found on the hard and soft palate.[9-12] Taste buds contain several types of cells, several of which are sensitive to taste stimuli during particular portions of their lives.[13] The anatomy of the tongue, papillae, and taste cells is shown in Plate 1. The three types of papillae and their distribution on the tongue can be seen there. Fungiform papillae are mushroom-shaped and distributed over the anterior dorsal two thirds of the tongue. About half of these papillae contain taste buds.[14-17] Foliate papillae consist of folds on the posterior lateral surface of the tongue. There are 3 to 8 of these papillae, each of which contains about 120 taste buds.[18] These papillae are difficult to observe in practice because extending the tongue to view them stretches the surface of the tongue and results in the folds becoming flat surfaces. Circumvallate papillae are raised mounds on the back of the tongue. Each of the 8 to 12 circumvallate papillae is surrounded by a moat which contains about 200 taste buds.[19] Fungiform papillae are innervated by the lingual branch of the seventh cranial nerve, the chorda tympani, and the foliate and circumvallate papillae are innervated by the lingual branch of the ninth cranial nerve, the glossopharyngeal nerve. These nerves carry information into the central nervous system where their pathways eventually lead to the gustatory neocortex. Relatively little is known about the central taste pathways and their function, although this area of study is now receiving considerable attention.[20-25]

B. TASTE SENSATIONS

Taste sensations are consciously experienced when neural signals originating at the receptor cells are relayed through their connecting neural pathways and are received by the appropriate area of the central nervous system. The initiation of this process occurs when a sapid substance interacts with the surface of a receptor cell. It is generally believed that the interaction of a sapid substance with the receptor cell results in electrical or conformational changes in the cell which ultimately lead to the release from the receptor of neurotransmitter chemicals in the vicinity of nerve fibers at the base of the receptors.[23,24] This process of converting the energy in the stimulus molecules, either chemical or electrical, into neural signals is called transduction. The transmitters released initiate signals in the nerve fibers near the base of the receptors which eventually reach the areas of the central nervous system responsible for processing taste information and result in the experience of taste. The entire process from arrival of stimulus material at the receptor surface until the actual sensation is experienced can take as little as 100 ms,[25] but varies considerably among different tastes as well as among different stimuli having the same taste.[26-32]

1. Number of Tastes

The word taste is used colloquially to designate any of a wide variety of sensations experienced in the oral cavity. In this usage, the word taste includes such sensations as texture, temperature, burning sensations associated with hot foods, and even olfactory sensations experienced retronasally. In order to systematically study sensations evoked by various stimuli, it is necessary to more systematically categorize sensations according to the sensory systems which mediate them. By this approach, tastes are designated as those sensations which are mediated by the taste receptor cells contained within taste buds. When care is taken to eliminate olfactory, tactile, and other sensations from consideration, it is found that there are very few different tastes. At present, evidence supports the existence of only four tastes: salty, sour, sweet, and bitter.[33,34] These data provide support for what has come to be known as the primary or basic taste position, that is, the position that these four qualities exhaustively represent taste quality. As a general rule (with exceptions easy to find), sugars taste sweet, alkalide halides taste salty, acids taste sour, and alkaloids taste bitter. However, the existence of only four tastes is controversial,[35-37] and arguments have been made for the existence of a taste continuum of many different taste sensations with salty, sour, sweet, and bitter representing only certain regions of the continuum. Detailed discussions and histories of this debate are available in several sources.[38-41]

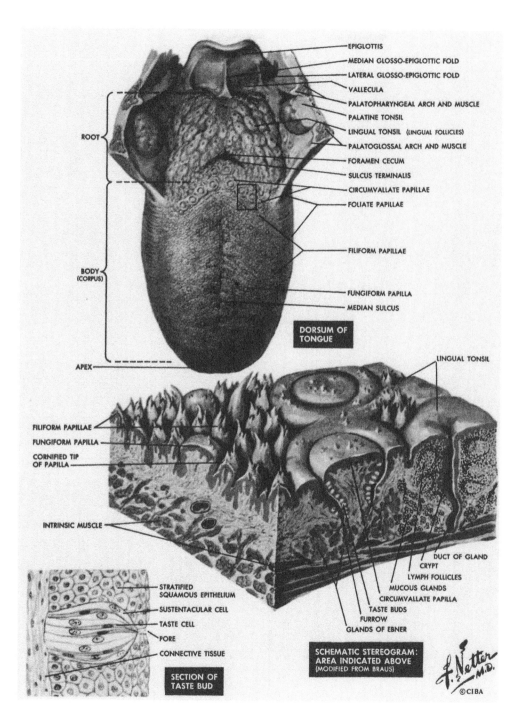

PLATE 1. Anatomy of the tongue showing the three types of taste papillae and the organization of taste receptor cells into a taste bud. (From Netter, F. H., *The CIBA Collection of Medical Illustrations*, CIBA Pharmaceutical Company, 1983. With permission.)

A related issue is whether taste is an analytic or synthetic sensory system.[39] An analytic system is one in which components of a mixture of stimuli remain individually identifiable. Audition is generally regarded as an analytic system because two tones of different frequency, when sounded together, can be individually heard and identified. In a synthetic system mixture, components blend together to create a new sensation which is qualitatively different from any of the individual components which are no longer directly available to experience. Vision is the classic example of a synthetic system, and color mixtures are generally given as examples of synthesis. Taste is widely thought to be analytic. The two facts most often given in support of this position are that taste mixtures do not produce new sensations in the same way that color mixtures do and that tastes do not cancel one another as do complementary colors.[40,43] Again, this question is somewhat controversial and it has been pointed out in several experiments that subjects in taste studies behave as if taste were somewhere between an analytic and synthetic system.[42,43]

Under proper conditions subjects can attend to a taste mixture as if it were a unitary sensation, while under other conditions they can analytically attend to single components of a complex mixture.[29] These latter findings are perhaps not at all surprising to experienced flavorists who routinely seem able to perceptually dissect a taste mixture into its components and also attend to the overall configuration of sensations that characterizes a particular mixture.

The analytic-synthetic debate is important beyond its theoretical implications. For example, if one believes that tastes in a mixture are not individually identifiable, and that new sensations different from the mixture components are actually experienced, then it makes no sense to ask for taste ratings of salty, sour, sweet, and bitter tastes. Instead, attention should be directed at identifying consistent quality labels for the various new sensations. Such attempts have been made, but no clear labels beyond the four traditional ones have yet been found.[44] On the other hand, when subjects are asked to rate the intensity of salty, sour, sweet, and bitter in mixture with each other[45,46] or as they occur in various foods,[47] they can do so easily and reliably.

C. TASTE INTENSITY
1. Thresholds

A detection threshold is the lowest concentration of a stimulus that can be detected. The recognition threshold is the lowest concentration that can be recognized as having a specific taste and is usually a bit higher than the detection threshold. Difference thresholds refer to the smallest change in concentration of a stimulus which can be just noticed. Detailed discussions of thresholds and how to obtain them are contained in Chapter 6. The present discussion is limited to some general issues about thresholds that have bearing on their interpretation.

Thresholds are often thought of as that concentration of stimulus which defines the lower limit of sensitivity of the sensory system. Concentrations below that level are assumed to not have any effect on the sensory system and therefore cannot be perceived. The intuitive appeal of this idea is obvious: there must be *some* absolute value below which the sensitivity of the sensory system does not permit detection. In actual fact, the concept of threshold is a statistical one. Thresholds are conventionally defined as that concentration which can be detected or recognized 50% of the time, or that difference in concentrations which can be resolved 50% of the time. However, there is no reason why an investigator may not choose some other more stringent criteria for defining threshold, say 75% detection. These statistical criteria are necessary because empirical data demonstrates detectability of stimuli does not follow a step function in which detection jumps from 0 to 100% at some specific value. Rather, data indicate that probability of detection increases gradually as concentration of a stimulus increases.[48]

In addition to the probabilistic nature of thresholds, their value is also dependent on the expectations brought to the threshold test by the subject.[49] For example, a very conservative subject may be less likely to report that he or she detects a sensation than a more liberal subject. Further, the subject's criterion for reporting a sensation may change during the course of a measurement session and even from trial to trial.[50] These problems are dealt with by signal detection methodology which is also discussed in Chapter 6.

Given the above considerations, it is still valuable to consider threshold as a useful concept which provides a rough indication of stimulus effectiveness or, conversely, subject sensitivity. In taste, bitter substances such as quinine tend to have low thresholds, being detectable at around $10^{-5} M$, while glucose (sweet) can be detected at about $10^{-1} M$ and NaCl (salty) at $10^{-2} M$; HCl (sour) is detectable at concentrations of about $10^{-3} M$. In general, a concentration change of around 20% is found to be the difference threshold in taste. These values are very approximate and will change as a function of many variables. For example, rinsing the mouth of saliva, and thus the NaCl contained in saliva, will lower the NaCl threshold about two log steps. Conversely, if the mouth is adapted to a given taste, the threshold for that taste will be increased to a value just higher than the adapting concentration.[51] Thresholds also vary with temperature, being lowest in the range of about 22 to 32°C.[52] An excellent source of approximate thresholds for a wide variety of taste and smell stimuli is the *ASTM Compilation of Odor and Taste Threshold Data*. This volume contains a listing of compounds which have been studied in threshold experiments along with a description of the procedure used to determine the threshold. Examining this catalogue provides a good idea of the type of variability one should expect in threshold data.

Thresholds are classically thought of as determinations made on individual subjects. Historically this has been the case. Examples of how thresholds are used with individual subjects are evident in the clinical literature where they are used as diagnostic tools in evaluation of the sensory systems.[53] However, there is no *a priori* reason why the computational procedures cannot be applied to populations as well as individuals. For example, applying threshold procedures to large populations can provide an estimate of what percentage of the population is likely to detect a product formulation change. This information can then be used to establish quality control limits.

2. Suprathreshold Intensity

The intensity of tastes increases as concentration of the tastant increases. The increase is not related to the threshold of the tastant. For example, the sweetness of sucrose increases with concentration faster than does the sweetness of Na-saccharin, but saccharin has a lower threshold. Thus, thresholds are not a good indicator of the relative intensity of a stimulus. In this regard, thresholds are best considered as indicators of stimulus effectiveness, i.e., a measure of how much of the stimulus is needed to reach a given sensation magnitude defined as threshold. In addition, rate of growth of sensation magnitude is not related to differential threshold, which is best considered as a measure of resolving power[54,55] of very small differences.

Since thresholds and rate of growth of sensation magnitude are not related, other indices are needed to compare compounds of different taste quality. One way to do this is by comparing their psychophysical functions. A psychophysical function is one which describes the change in sensation as a function of stimulus concentration. The growth in taste intensity with increases in stimulus concentration is often found to be very well described by a power function, $Y = cI^n$, where Y is the sensation magnitude, I is the stimulus concentration, and c is a scaling constant. The exponent n is the slope of the function plotted in log-log coordinates. The exponent provides a convenient index by which different tastes can be compared. For example, the exponent of the function describing the sweetness of sucrose is found to be greater than 1.0,[56] meaning that doubling the concentration of sucrose will result in an increase in sweetness which is more than doubled; Na-saccharin has been found to have a lower exponent, with 0.3 being a representative value.[57] This means that doubling the concentration of Na-saccharin will result only in about a 25% increase in sweetness intensity. Representative exponent values for other tastes are NaCl (salty), 1.2, and bitter (quinine) and sour (acids), 1.0.[58]

While power functions are certainly not the only way (and sometimes may not even be the preferred way) of describing taste intensity, they do reveal the important fact that taste intensity does not grow at the same rate for all tastes, and not even for all chemicals having the same taste quality. This fact will emerge no matter how the functions are mathematically described. From a practical standpoint, this is one of the most important properties of the human instrument because it describes the output of the system. The parameters of psychophysical functions tend to be only relatively constant and can vary considerably across experiments.[59] This fact indicates that not all variables affecting the output of humans as measuring devices have yet been identified.

D. SPATIAL DIFFERENCES IN SENSITIVITY

All areas of the tongue and palate that contain taste buds are sensitive to all tastes. This is not surprising because, although not all papillae contain taste buds, the ones that do are usually sensitive to more than one taste.[60] However, relative sensitivity does vary across the surface of the tongue and palate.[61] The area of the tongue containing foliate papillae (i.e., the lateral edges) is most sensitive to sour, while the front of the tongue has relatively low thresholds for salty, sweet, and bitter. However, the most sensitive area of the mouth to bitter is the soft palate, which has a threshold for quinine and bitter even lower than does the tip or back of the tongue. In addition, the slope of the bitter psychophysical function is steepest on the back of the tongue in the region of the circumvallate papillae, that is, bitterness increases rather quickly with stimulus concentration in that region. The low bitter threshold of the soft palate combined with the relatively steep psychophysical function for bitter of the circumvallate papillae account for the notable phenomenon of bitterness generally being much more apparent in the back of the mouth than in other regions.[61]

The size of the area of tongue stimulated by a taste affects the perceived intensity of the taste such that area stimulated multiplied by concentration is roughly equal to a constant taste intensity,[62] that is, as concentration decreases, area stimulated must increase to maintain a constant sensation magnitude. Some important consequences of this are that whole mouth tasting of samples results in a greater ability of subjects to discriminate intensities,[63] steeper psychophysical functions than tasting limited to smaller regions of the mouth,[59] and even that the rules governing how components of mixtures will add in intensity are different for whole-mouth tasting than for tasting limited to a specific tongue region.[64] The investigator looking for optimum sensitivity and discriminability from subjects should clearly encourage subjects to ensure thorough contact of all oral surfaces by flavor samples. Conversely, there may be occasions where it is desirable to limit tasting to specific tongue regions to, for example, exclude the influence of saliva. In those cases, the investigator should expect lower discriminability and shallower psychophysical functions.

E. ADAPTATION

Adaptation is the loss of sensitivity produced by continuous stimulation of a sensory system. The decreased sensitivity is specific to concentrations of stimuli equal to and less than that producing the adaptation. A common example of this is that everyone is adapted to the concentration of NaCl in their saliva. Concentrations at or below salivary levels of NaCl cannot be tasted, but higher concentrations can; if the mouth is rinsed of NaCl, then concentrations lower than salivary levels can be tasted.[51] Although taste adaptation can be complete for a given taste, achieving this is difficult,[65-67] presumably because salivary flow and tongue movements result in continuously changing stimulus concentrations at the receptors. However, high degrees of adaptation can be seen[65] as reduced perceived magnitude of a sensation over time. Recovery from taste adaptation is rapid and complete when the mouth is rinsed with water,[68] and can also occur more naturally by allowing the mouth to use its own natural means of clearing to take place, but this process is slower and the degree of recovery and mouth clearance is less certain.[69-71]

Adaptation is sometimes thought of as a sort of sensory fatigue, but this is not the case. The distinction between fatigue and adaptation will have very practical consequences on the design of sensory studies involving taste. Fatigue is a state of lessened capacity for work. This is clearly different from adaptation. In adaptation, the sensory system continues to respond to stimuli higher than the adapting stimuli in concentration; a fatigued system will be even less responsive to higher concentration. Consider, for example, muscular fatigue produced by repeatedly lifting a 20-lb weight. Muscles fatigued by that exercise will be even less capable of lifting a 40-lb weight. The taste system will, however, respond to concentrations higher than the adapting one. In addition, fatigue generally requires some relatively lengthy recovery time before the system can fully perform again. Adaptation is generally overcome rapidly and completely with thorough mouth rinsing. Managing this distinction in such a way as to decrease the possibility of adaptation and mental fatigue will allow the investigator to collect data from a considerable number of samples at one session, thus optimizing experimental efficiency.

If adaptation has occurred during the course of data collection, it can have very strong effects on the data obtained. One consequence involves the water taste.[72] After adapting to a given taste, water takes on a taste which depends on the quality of the adapting stimulus. For example, after adapting to salty tastes, the taste of water is bitter. Therefore, a salt-adapted subject will perceive any samples tasted in that state as less salty and more bitter because water in the sample will take on a bitter taste. The water taste has been extensively studied.[46,73] The general finding is that adapting to a given taste quality results in a decreased sensitivity to that quality (cross adaptation) and the addition of a specific water taste to subsequent samples. The addition of this water taste can be mistaken for an enhancement of the taste of a sample which already has some of that quality present because the water taste will add to the existing quality. This is not a true enhancement and it will disappear as soon as the water taste is eliminated. Investigators may be surprised to find that what they thought was a taste enhancement one day cannot be seen on another under different testing conditions. Another consequence of the water taste is the common observation that distilled water has a slight bitter taste. This is the result of the subject being adapted to saliva in the mouth. The water taste of distilled water will disappear if the taster first rinses his or her mouth thoroughly with the same distilled water before trying to taste it. Water tastes induced by 26 different tastants have been described and are shown in Figure 1.

Since the water taste will disappear with recovery from adaptation, it is very important that thorough recovery be insured between tasting samples. From a practical standing, mouth rinsing or allowing recovery by natural means between samples should be considered as important as taring a scale between weighing samples or rezeroing any other instrument to baseline before each use. Not rinsing is analogous to not changing a heavily used column on a chromatograph.

Exposure to tastants or other components in complex flavor samples can modify the taste system in other ways besides inducing adaptation. Sodium lauryl sulfate, a common surfactant found, for example, in toothpaste, has been shown to reduce sensitivity to sweet, salty, and bitter tastes, while adding a bitter taste to otherwise sour stimuli.[74] Some very common food constituents, such as sucrose, have also been found to alter taste sensitivity such as to enhance sour tastes independently of the water taste.[69] These effects are apparently due to alterations of receptor surfaces that may induce relatively long-lasting changes in sensitivity under some conditions.

F. MIXTURES

The intensity of individual tastes in a mixture depends on properties of the individual tastes it contains. In general, mixtures of stimuli having different qualities exhibit mixture suppres-

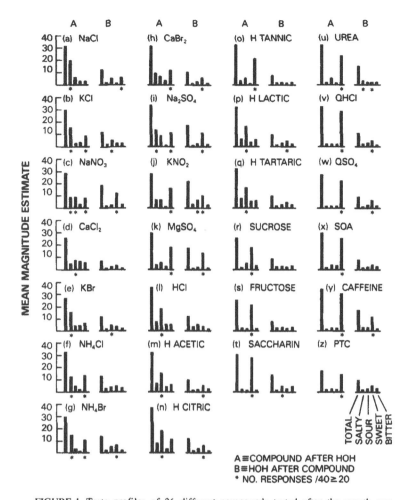

FIGURE 1. Taste profiles of 26 different compounds tasted after the mouth was adapted to distilled water (column A) and the taste of distilled water after the mouth was adapted to a solution of each of the 26 compounds (column B). Height of the bars indicates (from left to right) the total taste intensity and the intensity of the salty, sour, sweet, and bitter components of the total taste. Asterisk indicates the taste was used to describe the quality on at least half of the tasting trials. (From McBurney, D. H. and Shick, T. R., *Perception Psychophys.*, 10, 249, 1971. With permission.)

sion; that is, the components of the mixture taste less intense than they do at the same concentration as a single stimulus. The amount of suppression is related to the slope of the psychophysical functions of the mixture components and the number of tastes in the mixture as shown in Figure 2.[75] Those stimuli having the steepest slopes are suppressed less than those stimuli having relatively shallow slopes. This relationship to the slope of the psychophysical function (in log-log coordinates) also holds for mixtures of stimuli having similar tastes.[64] In this case, a related phenomenon called synergism can be demonstrated. Synergism occurs when the perceived intensity of a mixture exceeds the sum of intensities of the mixture components when they are tasted as individual stimuli. If the mixture components when tasted alone exhibit psychophysical functions with slopes greater than one, then mixing these stimuli will result in synergism; if the slopes of individual components are less than one when tasted alone, the mixture will exhibit compression.[64]

This relation of mixture intensities to the slope of the psychophysical function of the mixture components has some very important practical implications which make it important to understand what variables affect those slopes. It is well known that methods of tasting,[59] temperature,[76] and range of stimuli tested[77,78] all have pronounced effects on the slope of psy-

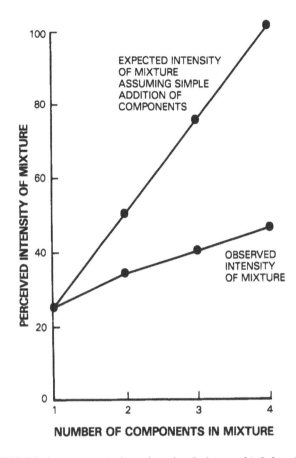

FIGURE 2. Average perceived taste intensity of mixtures of 1, 2, 3, and 4 taste components compared to the taste intensity expected assuming simple addition of the intensities of the components as tasted individually. The graph shows that, although total intensity of the mixtures increases, the increase is far short of additive. (From Bartoshuk, L. M., *Physiol. Behav.*, 14, 643, 1975. With permission.)

chophysical functions and therefore can be expected to affect the results of any tasting session. For example, Bartoshuk and Cleveland[64] found that a given mixture when tasted by a sip and spit procedure would exhibit synergism, but when flowed over the tongue would exhibit compression. This is because slopes of sip-and-spit procedures are higher than those obtained with flow stimulation. The apparent reasons for the difference are that in the sip-and-spit procedure more receptors are generally contacted by the samples and because flow procedures are often used with stimuli of different temperature than those of the sip-and-spit technique[76] simply because temperature can be more easily controlled with a flow system. It has been clearly demonstrated that slopes of psychophysical functions for taste decrease as temperature increases.[76]

An immediate concern to the product development person is that stimuli with the same taste quality will have different slopes. Knowledge of what these slopes are under the tasting conditions being used, or those expected to be used by consumers, will greatly facilitate efforts to determine optimal mixtures of similar tasting substances to deliver the desired effect. For example, if calorie reduction in a product is the desired goal, one way to achieve this is to reduce sugar content with artificial sweeteners. How much sweetener to add and still maintain the sweetness target can be determined by examination of slopes of various sugars and artificial sweeteners. However, since the slope of the sweetener function is related to how it interacts in mixture, changing sweetener will also result in changes in the other tastes present.

Some of the methodological variables affecting mixture tastes have already been discussed (i.e., slopes of psychophysical functions, tasting method, etc.). There are also properties of the taste system itself which affect mixtures. Some compression certainly can be accounted for by interactions of neurons in the very periphery of the taste system, i.e., in the tongue. For example, it has been found that taste stimuli initiate signals in neurons which proceed up the peripheral nervous system and into the brain, or central nervous system (CNS), and at the same time these signals are routed back toward the periphery by collateral branches of the neurons.[79,80] These returning signals then block the progress of other signals on their way to the CNS and thus are almost certainly involved in mixture suppression. This sort of effect is relatively common in the nervous system and is referred to as antidromic inhibition. It has also been found that some of the observed compression is apparently occurring in the CNS and not in the peripheral nervous system.[81,82]

The qualities of individual stimuli in mixtures can be recognized. However, the ability of tasters to attend to specific components of a mixture will vary with the type of tasting task used, and may also be affected by whether the instructions given to the tasters emphasize attending to individual components of the overall mixture configuration.[29] In order to facilitate accuracy and reliability of ratings, taste panel training ought to emphasize the subtle cues provided by the taste system[28] which can facilitate identifying and attending to single tastes in a complex mixture. These include differences in sensitivity to different tastes at various tongue loci, relative intensities of the tastes, the temporal properties of the various components as discussed below, as well as the instructional context established before each tasting session.[29] If unexpected results are obtained in a tasting session, or earlier results are not replicated, it is almost always due to unintended variability in one or more of the factors which affect mixture perception and because of the tasters' having difficulty perceptually dissecting the mixture. When confronted with such a situation, the prudent course of action is to list and then systematically examine the variables in the test and how they were controlled. When this is done the unintended influence can usually be found.

G. TEMPORAL ASPECTS

The taste system can extract useful information from stimuli in as little as 100 ms.[27,83] As the duration of the stimulus increases, more information is obtained. For example, it takes less time to detect the presence of a taste than to recognize what its quality is.[28] As a general rule, taste qualities can be rank ordered from fastest to slowest with regard to detection time as salty < sour < sweet < bitter,[28] and at least one study indicates that this ordering is likely maintained when stimuli of different qualities are mixed together.[29] There are exceptions to this generality; for example, the taste of Na-saccharin is detected faster than the taste of sucrose, both of which are primarily sweet.[28] Detection and recognition time also decrease as a function of stimulus concentration.[30-32] A formulator of oral products therefore has considerable leverage in determining which taste will be the first impression a subject in a rating session or a consumer will experience.

Taste stimuli also vary in temporal aspects besides detection and recognition time. Lee and Pangborn[84] have reviewed the literature and methods relating to taste intensity as a function of time of stimulation. Figure 3 graphically displays the type of information which they have pointed out is derivable from time-intensity (TI) studies in which subjects rate the intensity of stimuli at regular intervals, or continuously, over a period of time ranging up to a few minutes. The general finding of this work is that tastes have a characteristic time course including a brief initial rise time during which intensity increases, a peak intensity time, and an adaptation time during which the taste gradually decreases in intensity. As indicated in Figure 3, several different descriptive measures are derivable from these three basic parameters. As a general rule, different taste qualities exhibit different TI properties, and different stimuli with the same quality also vary,[84] again allowing considerable leeway in how the final sensory impression of a finished product can be constituted. An example of this is shown in

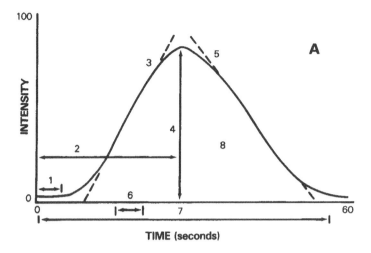

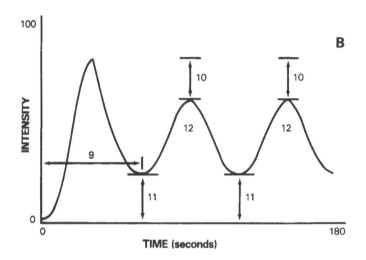

FIGURE 3. Information obtainable from time-intensity studies: (1) lag time, (2) time to maximum intensity, (3) rate of appearance, (4) maximum intensity, (5) rate of extinction or disappearance of taste, (6) events related to expectoration or swallowing, (7) total duration, (8) area under the curve, (9) time between introductions, (10) drop in maximum intensity relative to first introduction, (11) baseline shift, and (12) ratio of area under the curve to the area under the first introduction curve. (From Lee, W. E. and Pangborn, R. M., *Food Technol.*, 40, 71, 1986. With permission.)

Figure 4 from Larson-Powers and Pangborn.[85] These data demonstrate how the temporal properties of several product attributes were changed by substituting different sweeteners in an orange gelatin formulation, that is, not only do individual stimuli with the same or similar taste have different temporal properties, they have temporal mixture effects in that they change the temporal properties of other sensory aspects of mixtures in which they occur.

III. OLFACTION

A. ANATOMY OF THE OLFACTORY SYSTEM

Sensory receptors responsible for olfaction are located in the olfactory epithelium at the top of the nasal passages (Figure 5). Odorous molecules can reach this area either through the

ORANGE GELATIN

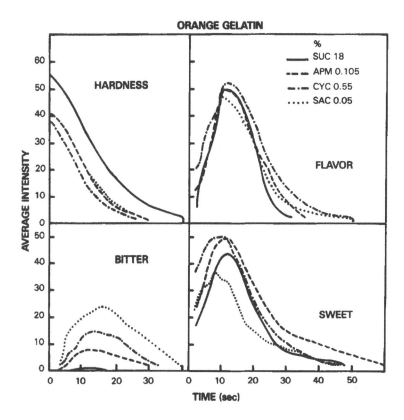

FIGURE 4. Changes in the shape of time-intensitiy curves resulting from the substitution of four different sweeteners in orange gelatin; Suc = sucrose, APM = aspartame, CYC = cyclamate, and SAC = saccharin. (From Larson-Powers, N. and Pangborn, R. M., *J. Food Sci.*, 43, 41, 1978. With permission.)

anterior nares via breathing or sniffing or through the posterior nares as when eating. The nasal passages and turbinate bones located there provide for a circuitous route by which inspired air reaches the area of the epithelium. Consequently, only a relatively small percentage of inhaled air actually comes into close proximity to the olfactory receptors. Depending upon flow rate, the amount of air reaching the epithelium is estimated at between 5 and 20%.[4] However, increased effort of sniffing (which increases flow rate) does not appear to influence the intensity of an odorant, although mechanically increasing flow into the nostrils using an olfactometer does seem to increase perceived intensity.[86-88]

The receptor cells themselves are embedded in the epithelium along with supporting and other cells. As with taste receptor cells, the olfactory receptors continuously turn over, with regeneration, maturing, and degeneration taking place in about 10 d.[89,90] The cells each have a number of cilia which project into a mucous layer covering the epithelium. These cilia are presumably the region where molecules interact with the receptors to initiate sensory signals.[91] The mucous covering serves to collect and concentrate molecules from air flowing over it, resulting in an increased odorant concentration at the receptors.[92] The other end of the olfactory receptors, a short axon, extends through a thin bone at the base of the skull, the cribiform plate, to contact second-order cells in the olfactory bulb. These cells begin the pathway which transmits olfactory signals to numerous regions in the brain.[18]

B. OLFACTORY SENSATIONS

As in taste, olfactory sensations arise when stimulus molecules interact with receptors and the consequent electrical or conformational changes in the cell initiate a neural signal which

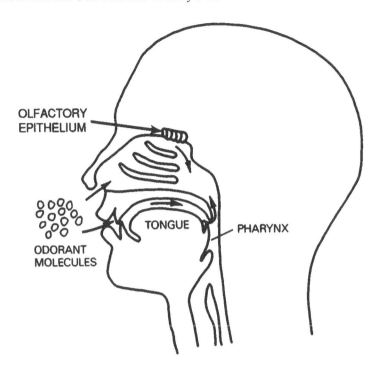

FIGURE 5. Anatomy of the olfactory system showing location of the olfactory epithelium which contains the olfactory receptor cells and the nasal and oropharyngeal routes by which odorous molecules may reach them.

is transmitted to the central nervous system.[18] Unlike taste, there is an incredibly wide variety of different odors. The main component of flavor is olfactory, not gustatory, as is commonly believed and colloquially expressed by such phrases as "the taste of food". It can be easily demonstrated that the sensations arising from food in the mouth are olfactory, and not taste, by holding one's nose closed before placing food in the mouth. Under these conditions, some or all of the four tastes will be experienced. Upon releasing the nose and breathing, the olfactory components will be experienced as the flavor of the food. A second thing that will be noticed is that the locus of the olfactory sensations seems to be the mouth. This very common and compelling illusion has been systematically investigated[93] and is likely mediated through the tactile sensation of food in the mouth, and may be physiologically determined by the convergence of olfactory, gustatory, and tactile neural information[93,94] in the CNS.

1. Number of Odors and Odor Classification

Even a brief perusal of the flavor and perfumery literature will give an immediate feeling for the number of odorous compounds, and the number of different odors, that exist. Odors are commonly classified by the object which produces them, e.g., fruity, grassy, etc. There have been numerous attempts to classify odors into primary categories such that every odor could fit into one category. Detailed reviews of these attempts are available[4,95,96] and will not be discussed here in depth. An example of a recent attempt at classification is that of Amoore,[97] who proposed seven classes which he called ethereal, floral, pepperminty, camphoraceous, musky, pungent, and putrid. Attempts at classification have been hindered by the lack of any identified underlying physical continuum along which odors can be seen to vary systematically in quality in a manner analogous to the variation of color with wavelength of visible light. Other difficulties have been the idiosyncratic nature of odor naming, i.e., the same compound may be called something quite different by different individuals.[98,99] Attempts to avoid this problem by using multidimensional scaling techniques, which circumvent language and vocabulary limitations by asking subjects to classify odors according to how similar they are to each other,[100] have also encountered difficulties.[4] For example, it has been very difficult to eliminate the hedonic component from these judgments.[101] In addition, groups of compounds which have been classified as similar in one study may be found to actually contain subgroups of their own in studies which focus only on those compounds. That is, the number of groupings found may depend on the number and qualities of odors studied.[4]

C. SYNTHESIS AND ANALYSIS IN OLFACTION

It is very clear that simple mixtures of odors can be analyzed into their components. It is also very obvious that complex mixtures are generally not analyzed and that the mixture components combine in a fashion that results in a sensation that is different from the qualities of the components. This is the accepted definition of a synthetic sensory system.[39] Examples of olfactory synthesis are common in everyday life. Coffee aroma is contributed to by several hundreds of compounds, a great many of which do not smell anything like coffee. Numerous examples of the same phenomenon can be found in *Fenaroli's Handbook of Flavor Ingredients*.[102]

It seems likely that the ability to analyze complex odors can be learned. Although the person on the street may typically be able to analyze only very simple mixtures, experienced flavorist's seem able to analyze mixtures that contain many more components. Almost certainly, some of this ability is not true analysis, but rather a reflection of the flavorists knowledge of which odors or chemicals are needed to create a certain sensation. On the other hand, the practical success of the various flavor training techniques argues strongly that the individual's ability to analyze can be enhanced,[47] but even here at least part of the training effect can likely be accounted for by the acquired ability to consistently apply a label to odors that could always be appreciated, but not identified,[103,104] i.e., training aids in the consistent use of odor labels.

D. THRESHOLDS AND INTENSITY

The olfactory system is exquisitely sensitive. Olfactory thresholds are frequently expressed in terms of parts per billion (ppb) of the compound in air,[102] and experienced flavor chemists know that the sensitivity of the nose surpasses that of even the most sensitive chromatograph. This sensitivity is in part contributed to by the ability of the olfactory mucosa to collect and concentrate odorous molecules at the receptors. Resolving power of the olfactory system is also quite good with differences in odorant concentrations in the range of about 5 to 10% being detectable (difference threshold) despite the many sources of variability that can enter into olfactory threshold determination experiments.[4,105,106]

When increases in odor intensity are plotted as a function of odorant concentration in log-log plots (power functions), the slopes are found to be somewhat lower than those for taste with values in the range of 0.2 to 0.4 being common. Of course, exceptions can easily be found with exponents ranging up to 1.0 or greater,[107] and Engen suggests that a representative value is about 0.6.[4] The shallow slopes, combined with low thresholds, results in olfactory sensitivity to stimulus concentrations that vary over several orders of magnitude.

Slopes of psychophysical functions for odors are not constants and can be expected to vary with experimental conditions in ways that may not be anticipated.[108] It is probably not a good idea to reject a piece of data because the slopes of two studies did not reproduce; such situations are likely to provide practical knowledge about the products being studied or techniques being used in the studies if the differences across testing occasions can be identified; that is, instead of indicating a lack of reliability, such situations probably point to an as-yet unidentified variable which impacts on results in a way not previously known. Hyman[108] has provided a review of variables which affect the exponent of psychophysical odor functions and has presented a summary table of representative exponent values for several different odorants.

E. ADAPTATION

Continuous exposure to an odorant leads to a gradual reduction in its intensity to somewhat less than half of its original intensity,[109] but has not been observed to be complete in studies which have not led the subjects to expect it would be. Recovery from adaptation occurs within a few minutes.[109] As with taste, threshold concentration of an odorant is increased by adaptation to a value slightly higher than that of the adapting concentration.[110]

F. MIXTURES: MASKING AND COUNTERACTION

The two most typical odor mixture phenomena are masking and counteraction.[111] Masking refers to the modification of odorant quality and is usually of interest with regard to malodors, or in the case of flavors, could refer to an off flavor in a complex system. Counteraction is the reduction in perceived intensity of an odorant or olfactory flavor component. Discussions of both of these effects, as well as less common mixture phenomena, are found in a review by Cain,[111] and Figure 6 is a schematic of the types of odor interactions that can occur.

Masking is most easily achieved when the odor to be masked is of low perceptual intensity and the masking odor is relatively more complex and more intense. Apparently, the complexity of the masking agent results in a sort of perceptual ambiguity about whether or not certain other odors are present.[111] The most common form of counteraction is called compromise. Compromise results when the total intensity of a mixture lies somewhere between the intensities of the unmixed components.

Perhaps due to the richness of olfactory experience (the sheer number of different olfactory sensations, and the fact that an odor quality may change in intensity *and* quality as its stimulus increases in concentration), a detailed understanding of complex odor mixtures is not available. Successful steps have been taken in that direction and certainly represent a start toward a better understanding. The total intensity of a two-component mixture can be described by a vector summation model:[112,113]

$$Y_{ab} = (Y_a^2 + Y_b^2 + 2Y_aY_b\cos\alpha)^{0.5} \tag{1}$$

In this model, Y_{ab} represents the perceived intensity of a two component odor mixture. Y_a and Y_b represent the perceived intensities of the two components when smelled as single stimuli. The α parameter is thought of as an angle between two vectors, Y_a and Y_b. The angle between the vectors is proportional to the similarity between the odors. If $\alpha = 0°$, the odors are very similar and their intensities add completely; when $\alpha = 180°$, the odors are very dissimilar and the perceived intensity of their mixture is equal to the absolute difference between their perceived intensities as single stimuli. Considerable data have been collected to demonstrate the utility of this model for describing total intensity of binary mixtures and higher-order mixtures.[114] The model does not describe how individual components will affect one another in mixture. The vector model certainly provides a starting point for thinking about more complex mixtures, and the parameter α may be useful as an index for cataloging the similarities among stimuli.

G. TASTE-SMELL MIXTURES

The major determinant of the flavor of foods and beverages is the mixture of taste and smell. The taste of a food in the mouth is primarily the experience of these sensations and, even though we refer to it as a taste of food, olfaction is the larger contributor. The tactile stimulation of the food substance in the mouth and the simultaneous retronasal stimulation of the olfactory receptors results in the illusion of the entire sensation being in the mouth.[93,94] The extent to which an odor and taste will interact is dependent on the perceptual quality of the stimuli. For example, the sweet taste of sucrose is enhanced by the lemon-like odor of citral,

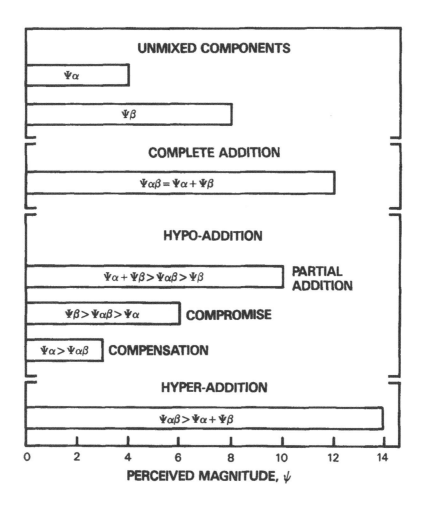

FIGURE 6. Possible effects of mixing two odorous compounds on the resulting perceived intensity of the mixture. Total intensity of the mixture may equal the sum of the intensities of the unmixed components or be less (hypoaddition) or greater (hyperaddition) than the sum of the unmixed intensities. (From Cain, W. S. and Drexler, M., *Ann. N.Y. Acad. Sci.*, 237, 427, 1974. With permission.)

but the salty taste of sodium chloride is not,[93] and other similar interactions have also been described.[115]

The important question from the perspective of chemometrics is how the blending of taste and smell into a flavor can be described. Several studies have found virtually complete additivity of taste and smell in the resultant flavor,[93,116] that is, overall flavor intensity was found to be about equal to overall intensity of the odorant plus the overall intensity of the tastant. These studies have generally required subjects to place the entire stimulus, containing both olfactory and gustatory components, in the mouth and then rate the overall intensity of smell and the overall intensity of taste. The sum of these estimates is then compared to estimates of overall flavor intensity and the result is complete additivity.

A different procedure which delivers the olfactory component directly to the nose and simultaneously delivers the gustatory component to the mouth has been found to result in less than total additivity,[117-119] that is, overall flavor intensity is less than the sum of the separate intensities of taste and smell. A simple model describes this result:

$$Y = k_s \text{ smell} + k_t \text{ taste} \tag{2}$$

where Y is the overall flavor intensity and k_s and k_t are correction factors to account for the difference between overall intensity and the sum of taste and smell intensities. This model implies that the overall intensity of a flavor containing no taste component would be judged to be less than judgments of smell alone and that the overall flavor intensity of a flavor containing no smell would be less than the judged taste intensity. This somewhat counterintuitive prediction has been found in experimental results.[119] Differences between overall intensity and estimates of taste when smell equals zero and of smell when taste equals zero have been used to empirically determine values for k_t and k_s, respectively.[119] It seems likely that these correction factors would be stimulus specific and thus could serve as the basis of models for various flavor systems. That overall flavor intensity is less than the sum of taste and smell, even if one of these sensations is not present, may indicate the operation of some cognitive factors in determining judged flavor intensity, or is perhaps related to task differences. Earlier work showing additivity or near additivity placed the entire stimulus (both taste and smell components) in the mouth and compared estimates of *overall* taste intensity plus *overall* smell intensity to *overall* flavor intensity. The work showing less complete additivity compared estimates of taste and smell to *overall* flavor intensity defined as "some combination of smell and taste" and delivered olfactory stimuli to the nose and gustatory components to the mouth.

IV. THE COMMON CHEMICAL SENSE

The term common chemical sense refers to sensations such as burning, tingling, biting, cooling, and so forth, which are associated with spicy or hot foods and such stimuli as menthol. These same sensations are often referred to as trigeminal sensations, after the trigeminal nerve which mediates their perception. Capsaicin is the common chemical stimulant in hot peppers, and CO_2 is often used as a nasal trigeminal stimulant and elicits a sensation of pungency.[120,121] Other trigeminal stimulants include horseradish, ginger, black pepper, vinegar, and several other spices. In flavor work, interest in common chemical sensitivity is restricted to the mucous membranes of the nose, mouth, and throat, but any mucous membrane will be sensitive to stimuli eliciting these sensations.

A. ANATOMY

Common chemical sensitivity is subtended by the trigeminal nerve which splits into several branches to innervate the tongue, oral mucosa, nasal mucosa, throat, and other areas of the head and face.[122] The endings of the individual trigeminal neurons are not associated with any organized type of receptor and are therefore called free nerve endings. These endings are very dense in the nasal and oral mucosa, accounting for up to 75% of all nerve endings in fungiform papillae.[123]

B. PERCEPTION

As noted above, trigeminal stimuli elicit sensations such as pungency, cooling, burning, etc. The characteristics of these sensations provide interesting contrasts to the taste and smell sensations more often considered in flavor. For example, trigeminal sensations greatly outlast the stimulus and may be perceptible for more than 10 min even with repeated mouth rinsing with plain water or water containing other taste stimuli.[124] Further, unlike taste and smell sensations which generally decrease in intensity with repeated presentation of the same stimulus, trigeminal sensations in both the oral cavity and nasal passages increase sharply with repeated stimulus presentations.[125,126] These two facts, prolonged sensation duration and increased intensity with repeated presentation, make psychophysical study of trigeminal sensations difficult. Sensory evaluations of products containing such stimuli are wisely limited to only one or a few tastings with lengthy recovery periods between samples. Trigeminal sensations also tend to have steeper psychophysical slopes than taste and odor stimuli[121] and have an increased latency of detection:[126] it takes longer for the sensation to develop after the stimulus is delivered than it does for taste or smell sensations to develop. However, trigeminal sensations show considerable temporal summation, i.e., the sensation grows in intensity with prolonged application of the stimulus.[126]

Both taste and smell interact with the common chemical sense in a fashion similar to mixture suppression. Odors are perceived as less intense in the presence of a trigeminal stimulant[127] and the reverse is also true, with similar relations holding for taste-common chemical sense mixtures.[124]

It is not clear that only one type of trigeminal pain-like sensation exists. For example, different chemical irritants are more or less effective as stimuli at different oral loci, are described differently by subjects,[120] and inhibit taste sensations to different degrees.[124] Further, the burning produced by the second presentation of an oral irritant is greater if the second presentation is with a stimulus different from the first.[125] This suggests that the two irritants are affecting different populations of neurons and the second stimulus, if different from the first, recruits these different neurons to contribute to the total sensation.[125] All of these data have been seen as suggesting the existence of qualitatively different trigeminal sensations.[124]

The presentation of both oral and nasal trigeminal stimuli elicits reflex responses. Nasal stimuli elicit a definite but transient interruption in breathing, referred to as apnea, upon inhalation.[127,128] Interestingly, the reflex is much reduced in cigarette smokers who have smoked a cigarette just before exposure.[128] This finding is in contrast to the earlier mentioned data showing that successive presentation of the same or different trigeminal stimulant is often associated with *increased* intensity of *sensation*.[125,126] Oral trigeminal stimuli elicit, among other things, increased salivary flow[130] and sweating in head and facial regions.[131] Both the nasal and oral reflexes apparently serve to reduce exposure to the irritants and remove their source.[122]

Trigeminal stimuli are initially aversive to virtually everyone and are generally rejected on first exposure. However, repeated exposure reverses this innate aversion[131-134] and preference for the sensation develops. This may in part be due to the enhancement of otherwise bland diets,[131] sensation seeking,[132-134] or other mechanisms. The preference development apparently has some cognitive or cultural components as well,[131] because animals do not develop similar preferences with repeated exposure.[135]

REFERENCES

1. **Barlow H. B. and Mullon J. D.,** *The Senses.* Cambridge University Press, Cambridge, 1982.
2. **Carterette, E. C. and Friedman, M. P., Eds.,** *Handbook of Perception,* Vol. 6A, Academic Press, New York, 1978.
3. **Darian-Smith, I., Ed.,** *Handbook of Physiology,* Vol. 3, American Physiological Society, Bethesda, MD, 1984.
4. **Engen, T.,** *The Perception of Odors.* Academic Press, New York, 1982.
5. **Marks, L. E.,** *Sensory Processes: The New Psychophysics,* Academic Press, New York, 1974.
6. **Uttal, W. R.,** *The Psychobiology of Sensory Coding,* Harper & Row, New York, 1983.
7. **Arvidson, K. and Friberg, U.,** Human taste: response and taste bud number in human fungiform papillae, *Science,* 209, 807, 1980.
8. **Miller, I. J. and Spangler, K. M.,** Taste bud distribution and innervation on the palate of the rat, *Chem. Senses,* 7, 99, 1982.

9. **Miller, I. J. and Smith, D. V.,** Quantitative taste bud distribution in the hamster, *Physiol. Behav.,* 32, 275, 1984.

10. **Nilsson, B.,** Taste acuity of the human palate. III. Studies with taste solutions on subjects in different age groups, *Acta Odontol. Scand.,* 37, 235, 1979.

11. **Kinnamon, J. C.,** Organization and innervation of taste buds, in *Neurobiology of Taste and Smell,* Finger, T. E. and Silver, W. L., Eds., John Wiley & Sons, New York, 1987.

12. **Arvidson, K.,** Scanning electron microscopy of fungiform papillae in the tongue of man and monkey, *Acta Otolaryngol.,* 81, 496, 1976.

13. **Paran, N., Mattern, C., and Henkin, R.,** The ultrastructure of the taste bud of the human fungiform papilla, *Cell Tissue Res.,* 161, 1, 1975.

14. **Mochizuki, Y.,** Studies on the papilla foliata of Japanese. II. The number of taste buds, *Okajimas Folea Anat. Jpn.,* 18, 355, 1939.

15. **Arey, L. B., Tremaine, M. J., and Monzingo, F. L.,** The numerical and topographical relations of taste buds to human circumvallate papillae throughout the lifespan, *Anat. Rec.,* 64, 9, 1935.

16. **Finger, T. E.,** Gustatory nuclei and pathways in the central nervous system, in *Neurobiology of Taste and Smell,* Finger, T. E. and Silver, W. L., Eds., John Wiley & Sons, New York, 1987.

17. **Hanamori, T. and Smith, D. V.,** Central projections of the hamster superior laryngeal nerve, *Brain Res. Bull.,* 16, 271, 1986.

18. **Norgren, R.,** Central neural mechanisms of taste, in *Handbook of Physiology,* Vol. 3, Darian-Smith, I., Ed., American Physiological Society, Bethesda, Maryland, 1984.

19. **Smith, D. V., Travers, J. B., and Van Buskirk, R. L.,** Brainstem correlates of gustatory similarity in the hamster, *Brain Res. Bull.,* 4, 359, 1979.

20. **Yamamoto, T.,** Taste responses of cortical neurons, *Prog., Neurol. Biol.,* 23, 273, 1984.

21. **Yamamoto, T., Azuma, S., and Kawamura, Y.,** Functional relations between the cortical gustatory area and the amygdala: electrophysiological and behavioral studies in rats, *Exp. Brain Res.,* 56, 23, 1984.

22. **Yamamoto, T., Matsuo, R., and Kawamura, Y.,** Localization of cortical gustatory area in rats and its role in taste discrimination, *J. Neurophysiol.,* 44, 440, 1980.

23. **Teeter, J. H. and Brand, J. G.,** Peripheral mechanisms of gustation: physiology and biochemistry, in *Neurobiology of Taste and Smell,* Finger, T. E. and Finger, W. L., Eds., John Wiley & Sons, New York, 1987.

24. **Teeter, J., Funakoshi, M., Kurihara, K., Roper, S., Sato, T., and Tonosaki, K.,** Generation of the taste cell potential, *Chem. Senses,* 12, 215, 1987.

25. **Halpern, B. P. and Tapper, D. N.,** Taste stimuli: quality coding time, *Science,* 171, 1256, 1971.

26. **Kelling, S. T. and Halpern, B. P.,** Taste judgments and gustatory stimulus duration: simple reaction times, *Chem. Senses,* 12, 543, 1987.

27. **Kelling, S. T. and Halpern, B. P.,** Taste flashes: reactions times, intensity, and quality, *Science,* 219, 412, 1983.

28. **Kuznicki, J. T. and Turner, L. S.,** Reaction time in the perceptual processing of taste quality, *Chem. Senses,* 11, 183, 1986.

29. **Kuznicki, J. T. and Turner, L. S.,** Temporal dissociation of taste mixture components, *Chem. Senses,* 13, 45, 1988.

30. **Yamamoto, T. and Kawamura, Y.,** Gustatory reaction time in human adults, *Physiol. Behav.,* 26, 715, 1981.

31. **Yamamoto, T. and Kawamura, Y.,** Gustatory reaction time to various salt solutions in human adults, *Physiol. Behav.,* 32, 49, 1984.

32. **Yamamoto, T., Kato, T., Matsuo, R., Araie, N., Azuma, S., and Kawamura, Y.,** Gustatory reaction time under variable stimulus parameters in human adults, *Physiol. Behav.,* 29, 79, 1982.

33. **McBurney, D. H.,** Are there primary tastes for man?, *Chem. Senses Flav.,* 1, 17, 1974.

34. **McBurney, D. H. and Gent, J. F.,** On the nature of taste qualities, *Psychol. Bull.,* 86, 151, 1979.

35. **Erickson, R. P. and Covey, E.,** On the singularity of taste sensations: what is a primary taste?, *Physiol. Behav.,* 25, 527, 1980.

36. **Erickson, R. P., Covey, E., and Doetsch, G. S.,** Neuron and stimulus typologies in the rat gustatory system, *Brain Res.,* 196, 513, 1980.

37. **Schiffman, S. S. and Erickson, R. P.,** A psychophysical model of gustatory quality, *Physiol. Behav.,* 7, 617, 1971.

38. **Bartoshuk, L. M.,** History of taste research, in *Handbook of Perception,* Vol. 6, Carterette, E. C. and Friedman, M. P., Eds., Academic Press, New York, 1978.

39. **Erickson, R. P.,** Stimulus coding in topographic and non-topographic afferent modalities: on the significance of the activity of individual sensory neurons, *Psychol. Rev.,* 75, 447, 1968.

40. **Erickson, R. P.,** Ohrwall, Henning, and von Skramlik; the foundations of the four primary position in taste, *Neurosci. Biobehav. Rev.,* 8, 105, 1984.

41. **McBurney, D. H.,** Taste, smell, and flavor terminology: taking the confusion out of fusion, in *Clinical Measurement of Taste and Smell,* Meiselman, H. L. and Rivlin, R. S., Eds., Macmillan, New York, 1986.

42. **Kuznicki, J. T.,** Space and time separation of taste mixture components, *Chem. Senses,* 7, 39, 1982.

43. **Scott, T. R. and Chang, F. T.,** The state of gustatory neural coding, *Chem. Senses,* 8, 297, 1984.

44. **Ishii, R. and O'Mahoney, M.,** Taste sorting and naming: can taste concepts be misrepresented by traditional psychophysical labeling systems?, *Chem. Senses,* 12, 37, 1987.

45. **Smith, D. V. and McBurney, D. H.,** Gustatory cross-adaptation: does a single mechanism code the salty taste?, *J. Exp. Psych.,* 80, 101, 1969.

46. **McBurney, D. H. and Shick, T. R.,** Taste and water tastes of twenty-six compounds for man, *Percept Psychophys.,* 10, 249, 1971.

47. **Meilgard, M., Civille, G. V., and Carr, B. T.,** *Sensory Evaluation Techniques,* Vol. 2, CRC Press, Boca Raton, FL, 1987.

48. **Corso, J. F.,** A theoretico-historical review of the threshold concept, *Psychol. Bull.,* 60, 356, 1963.

49. **Green, D. M. and Swets, J. A.,** *Signal Detection Theory and Psychophysics,* John Wiley & Sons, New York, 1966.

50. **Treisman, M. and Williams, T. C.,** A theory of criterion setting with an application to sequential dependencies, *Psychol. Rev.,* 91, 68, 1984.

51. **McBurney, D. H. and Pfaffmann, C.,** Gustatory adaptation to saliva and sodium chloride, *J. Exp. Psychol.,* 65, 523, 1963.

52. **McBurney, D. H., Collings, V. B., and Glanz, L. M.,** Temperature dependence of human taste response, *Physiol. Behav.,* 11, 89, 1973.

53. **Engen, T.,** Classical psychophysics: humans as sensors, in *Clinical Measurement of Taste and Smell,* Meiselman, H. L. and Rivlin, R. S., Eds., Macmillan, New York, 1986.

54. **Cain, W. S.,** Odor magnitude: coarse vs. fine grain, *Percept. Psychophys.,* 22, 545, 1977.

55. **Sauvageot, F.,** Differential threshold and exponent of the power function in the chemical senses, *Chem. Senses,* 12, 537, 1987.

56. **Moskowitz, H. R.,** Ratio scales of sugar sweetness, *Percept. Psychophys.,* 7, 315, 1970.

57. **Moskowitz, H. R.,** Sweetness and intensity of artificial sweeteners, *Percept. Psychophys.,* 8, 40, 1970.

58. **Moskowitz, H. R.,** Scales of Intensity for Single and Compound Tastes, dissertation, Harvard University, Cambridge, MA, 1968.

59. **Meiselman, H. L.,** Human taste perception, *Crit. Rev. Food Technol.,* April 89, 1972.

60. **Kuznicki, J. T. and Cardello, A. V.,** Psychophysics of single taste papillae, in *Clinical Measurement of Taste and Smell,* Meiselman, H. L. and Rivlin, R. S., Eds., Macmillan, New York, 1986.

61. **Collings, V. B.,** Human taste response as a function of locus of stimulation on the tongue and soft palate, *Percept. Psychophys.,* 16, 169, 1974.

62. **Smith, D. V.,** Taste intensity as a function of area and concentration: differentiation between compounds, *J. Exp. Psychol.,* 87, 163, 1971.

63. **Lawless, H. and Skinner, E. Z.,** The duration and perceived intensity of sucrose taste, *Percept. Psychophys.,* 25, 180, 1979.

64. **Bartoshuk, L. M. and Cleveland, C. T.,** Mixtures of substances with similar tastes: a test of a psychophysical model of taste mixture interactions, *Sen. Proc.,* 1, 177, 1977.

65. **Gent, J. F. and McBurney, D. H.,** Time course of gustatory adaptation, *Percept. Psychophys.,* 23, 171, 1978.

66. **DuBose, C. N., Meiselman, H. L., Hunt, D. A., and Waterman, D.,** Incomplete taste adaptation to different concentrations of salt and sugar solutions, *Percept. Psychophys.,* 21, 183, 1977.

67. **Meiselman, H. L. and Buffington, C.,** Effects of filter paper stimulus application on gustatory adaptation, *Chem. Senses,* 5, 273, 1980.

68. **Hahn, H.,** Die adaptation des geschmackssinnes, *Z. Sinnesphysiol.,* 65, 105, 1934, cited in **Mcburney, D. H.,** Taste and olfaction: sensory discrimination, in *Handbook of Physiology,* Adrian-Smith, I., Ed., The American Physiological Society, Bethesda, MD, 1984.

69. **Kuznicki, J. T. and McCutcheon, N. B.,** Cross-enhancement of the sour taste on single human taste papillae, *J. Exp. Psychol. Gen.,* 108, 68, 1979.

70. **Meiselman, H. L. and Halpern, B. P.,** Effects of *Gymnema sylvestre* on complex tastes elicited by amino acids and sucrose, *Physiol. Behav.,* 5, 1379, 1970.

71. **Riskey, D. R., Desor, J. A., and Vellucci, D.,** Effects of gymnemic acid concentration and time since exposure on intensity of simple tastes: a test of the biphasic model for the action of gymnemic acid, *Chem. Senses,* 7, 143, 1982.

72. **Bartoshuk, L. M.,** Water taste in man, *Percept. Psychophys.,* 3, 69, 1968.

73. **McBurney, D. H. and Bartoshuk, L. M.,** Interactions between stimuli with different taste qualities, *Physiol. Behav.,* 10, 1101, 1973.

74. **DeSimone, J. A., Heck, G. L., and Bartoshuk, L. M.,** Surface active taste modifiers: a comparison of the physical and psychophysical properties of gymnemic acid and sodium lauryl sulfate, *Chem. Senses,* 5, 317, 1980.

75. **Bartoshuk, L. M.,** Taste mixtures: is mixture suppression related to compression?, *Physiol. Behav.,* 14, 643, 1975.

76. **Bartoshuk, L. M., Rennert, K., Rodin, J., and Stevens, J. C.,** Effects of temperature on the perceived sweetness of sucrose, *Physiol. Behav.*, 28, 905, 1982.

77. **Poulton, E. C.,** The new psychophysics: six models for magnitude estimation, *Psychol. Bull.*, 69, 1, 1968.

78. **Teghtsoonian, R. and Teghtsoonian, M.,** Range and regression effects in magnitude scaling, *Percept. Psychophys.*, 24, 305, 1978.

79. **Miller, I. J.,** Peripheral interactions among single papilla inputs to gustatory nerve fibers, *J. Gen. Physiol.*, 57, 1, 1971.

80. **Rapuzzi, G. and Casella, C.,** Innervation of the fungiform papillae in the frog tongue, *J. Neurophysiol.*, 28, 154, 1965.

81. **Kroeze, J. H. A.,** After repetitive sucrose stimulation saltiness suppression in NaCl-sucrose mixtures is diminished: implications for a central mixture suppression mechanism, *Chem. Senses*, 7, 81, 1982.

82. **Kroeze, J. H. A. and Bartoshuk, L. M.,** Bitterness suppression as revealed by split tongue taste stimulation in humans, *Physiol. Behav.*, 35, 779, 1985.

83. **McBurney, D. H.,** Temporal properties of the human taste system, *Sens. Proc.*, 1, 150, 1976.

84. **Lee, W. E. and Pangborn, R. M.,** Time-intensity: the temporal aspects of sensory perception, *Food Technol.*, 40, 71, 1986.

85. **Larson-Powers, N. and Pangborn, R. M.,** Paired comparison and time-intensity measurements of the sensory properties of beverages and gelatins containing sucrose or synthetic sweeteners, *J. Food Sci.*, 43, 41, 1978.

86. **Rehn, T.,** Perceived odor intensity as a function of airflow through the nose, *Sens. Proc.*, 2, 198, 1978.

87. **Teghtsoonian, M., Teghtsoonian, R., Berglund, B., and Berglund, U.,** Comment on "Perceived odor intensity as a function of airflow through the nose", *Sens. Proc.*, 3, 204, 1979.

88. **Teghtsoonian, R., Teghtsoonian, M., Berglund, B., and Berglund, U.,** Invariance of odor strength with sniff vigor: an olfactory analogue to size constancy, *J. Exp. Psychol.: Hum. Percept. Perform.*, 4, 144, 1978.

89. **Costanzo, R. M. and Graziadei, P. P. C.,** A quantitative analysis of changes in the olfactory epithelium following bulbectomy in the hamster, *J. Comp. Neurol.*, 215, 370, 1983.

90. **Costanzo, R. M. and Graziadei, P. P. C.,** Development and plasticity of the olfactory system, in *Neurobiology of Taste and Smell*, Finger, T. E. and Silver, W. L., Eds., John Wiley & Sons, New York, 1987.

91. **Getchell, T. V. and Getchell, M. L.,** Peripheral mechanisms of olfaction: iochemistry and neurophysiology, in *Neurobiology of Taste and Smell*, Finger, T. E. and Silver, W. L. Eds., John Wiley & Sons, New York, 1987.

92. **Beets, M. G. J.,** The molecular parameters of olfactory response, *Pharmacol. Rev.*, 22, 1, 1970.

93. **Murphy, C. and Cain, W. S.,** Taste and olfaction: independence vs. interaction, *Physiol. Behav.*, 24, 601, 1980.

94. **Van Buskirk, R. L. and Erickson, R. P.,** Odorant responses in taste neurons of the rat NTS, *Brain Res.*, 135, 287, 1977.

95. **Moncrieff, R. W.,** *The Chemical Senses*, Leonard Hill, London, 1967.

96. **Cain, W. S.,** History of research on smell, in *Handbook of Perception*, Vol. 6A, Carterette, E. C. and Friedman, M. P., Eds., Academic Press, New York, 1978.

97. **Amoore, J. E.,** *Molecular Basis of Odor*, Charles C Thomas, Springfield, IL, 1970.

98. **Harper, R., Bate-Smith, E. C., and Land, D. G.,** *Odour Description and Odour Classification*, J. & A. Churchill, London, 1968.

99. **Engen, T. and Ross, B. M.,** Long term memory of odors with and without verbal descriptions, *J. Exp. Psychol.*, 100, 221, 1973.

100. **Schiffman, S. S., Robinson, D. E., and Erickson, R. P.,** Multidimensional scaling of odorants: examination of psychological and physiochemical dimensions, *Chem. Senses Flavor*, 2, 375, 1977.

101. **Schiffman, S. S.,** Preference: a multidimensional concept, in *Preference Behavior and Chemoreception*, Kroeze, J. H. A., Ed., IRL Press, London, 1979.

102. **Furia, T. E. and Bellanca, N.,** Eds., *Fenaroli's Handbook of Flavor Ingredients*, 2nd ed., CRC Press, Boca Raton, FL, 1975.

103. **Cain, W. S.,** To know with the nose: keys to odor identification, *Science*, 203, 467, 1979.

104. **Desor, J. A. and Beauchamp, G. K.,** The human capacity to transmit olfactory information, *Percept. Psychophys.*, 16, 551, 1974.

105. **Cain, W. S.,** Odor magnitude: coarse vs. fine grain, *Percept. Psychophys.*, 22, 545, 1977.

106. **Cain, W. S.,** Differential sensitivity for smell: "noise" at the nose, *Science*, 195, 796, 1977.

107. **Laing, D. G., Panhuber, H., and Baxter, R. I.,** Olfactory properties of amines and n-butanol, *Chem. Senses Flavor*, 3, 149, 1978.

108. **Hyman, A. N.,** Factors influencing the psychophysical function for odor intensity, *Sens. Proc.*, 1, 273, 1977.

109. **Ekman, G., Berglund, B., Berglund, V., and Lindvall, T.,** Perceived intensity of odor as a function of time of adaptation, *Scand. J. Psychol.*, 8, 177, 1967.

110. **Engen, T.,** Olfactory psychophysics, in *Handbook Sensory Physiology*, Beidler, L. M., Ed., Vol. 4, Part 1, Springer-Verlag, New York, 1971.

111. **Cain, W. S. and Drexler, M.,** Scope and evaluation of odor counteraction and masking, *Ann. N.Y. Acad. Sci.,* 237, 427, 1974.
112. **Berglund, B., Berglund, U., Lindvall, T., and Svensson, L. T.,** A quantitative principle of perceived intensity summation in odor mixtures, *J. Exp. Psychol.,* 100, 29, 1973.
113. **Berglund, B., Berglund, U., and Lindvall, T.,** Psychological processing of odor mixtures, *Psychol. Rev.,* 83, 432, 1976.
114. **Moskowitz, H. R.,** Utility of the vector model for higher-order mixtures: a correction, *Sens. Proc.,* 3, 366, 1979.
115. **Frank, R. A. and Byram, J.,** Taste-smell interactions are tastant and odorant dependent, *Chem. Senses,* 13, 445, 1988.
116. **Murphy, C., Cain, W. S., and Bartoshuk, L. M.,** Mutual action of taste and olfaction, *Sens. Process.,* 1, 204, 1977.
117. **Burdach, K., J., Kroeze, J. H. A., and Koster, E. P.,** Nasal, retronasal, and gustatory perception: an experimental comparison, *Percept. Psychophys.,* 36, 205, 1984.
118. **Enns, M. P. and Hornung, D. E.,** Contributions of smell and taste to overall intensity, *Chem. Senses,* 10, 357, 1985.
119. **Hornung, D. E. and Enns, M. E.,** The contributions of smell and taste to overall intensity: a model, *Percept. Psychophys.,* 39, 385, 1986.
120. **Lawless, H. T. and Stevens, D. A.,** Responses by humans to oral chemical irritants as a function of locus of stimulation, *Percept. Psychophys.,* 43, 72, 1988.
121. **Cometto-Muniz, J. E. and Noriega, G.,** Gender differences in the perception of pungency, *Physiol. Behav.,* 34, 385, 1985.
122. **Silver, W. L.,** The common chemical sense, in *Neurobiology of Taste and Smell,* Finger, T. E. and Silver, W. L., Eds., John Wiley & Sons, New York, 1987.
123. **Farbman, A. I. and Hellekant, G.,** Quantitative analyses of the fiber population in rat chorda tympani nerves and fungiform papillae, *Am. J. Anat.,* 153, 509, 1978.
124. **Stevens, D. A. and Lawless, H. T.,** Putting out the fire: effects of tastants on oral chemical irritation, *Percept. Psychophys.,* 39, 346, 1986.
125. **Stevens, D. A. and Lawless, H. T.,** Enhancement of responses to sequential presentation of oral chemical irritants, *Physiol. Behav.,* 39, 63, 1987.
126. **Cain, W. S.,** Olfaction and the common chemical sense: some psychophysical contrasts, *Sens. Proc.,* 1, 57, 1976; **Cain, W. S. and Murphy, C.,** Interaction between chemoreceptive modalities of odour and irritation, *Nature (London),* 284, 255, 1980.
127. **Cometto-Muniz, J. E. and Cain, W. S.,** Perception of nasal pungency in smokers and non-smokers, *Physiol. Behav.,* 29, 727, 1982.
128. **Stevens, J. C. and Cain, W. S.,** Aging and the perception of nasal irritation, *Physiol. Behav.,* 37, 323, 1986.
129. **Lawless, H. T.,** Oral chemical irritation: psychophysical properties, *Chem. Senses,* 9, 143, 1984.
130. **Lee, T. S.,** Physiological gustatory sweating in a warm climate, *J. Physiol.,* 124, 528, 1954.
131. **Rozin, P.,** The use of characteristic flavorings in human culinary practice, in *Flavor: Its Chemical, Behavioral, and Commercial Aspects,* Westview Press, Boulder, CO, 1977.
132. **Rozin, P. and Schiller, D.,** The nature and acquisition of a preference for chili pepper by humans, *Motivation Emotion,* 4, 77, 1980.
133. **Rozin, P., Mark, M., and Schiller, D.,** The role of desensitization to capsaicin in chili pepper ingestion and preference, *Chem. Senses,* 6, 23, 1981.
134. **Rozin, P., Ebert, L., and Schull, J.,** Some like it hot: a temporal analysis of hedonic responses to chili pepper, *Appetite,* 3, 13, 1982.
135. **Rozin, P., Gruss, L., and Berk, G.,** Reversal of innate aversions: attempts to induce a preference for chili peppers in rats, *J. Comp. Physiol. Psychol.,* 93, 1001, 1979.

Chapter 6

MEASURING SENSITIVITY

I. PSYCHOPHYSICS: THE BASIS OF SENSORY EVALUATION

Psychophysics is the science of the relationship between energy and the sensations it produces. In the consumer products industry, psychophysics is referred to as sensory evaluation. The subject matter, especially the methods, are the same, with the exception that industrial applications tend to be quite specific relative to the sometimes more general academic problems approached by psychophysics. For example, the objects of sensory (psychophysical) studies in the food industry are foods. This specificity of application sometimes leads to the belief of a fundamental difference between psychophysics and sensory evaluation. This viewpoint regards sensory evaluation as being concerned with properties of certain items like foods, beverages, paper products, etc., while psychophysics is concerned with properties of the humans who are used as measuring instruments. According to this latter view, the items studied are only there as a means to probe the human instrument for knowledge concerning its operating characteristics, and the data obtained are not directly useful in the understanding of sensory properties of the items per se.

Taken in their extremes, neither of these positions is wholly correct. While the interest of the practitioner may range anywhere between properties of the human to properties of the physical items being studied, it is well known that human input and physical energy interact to produce the data obtained, and one cannot be wholly understood in the absence of knowledge about the other. A more reasonable position is therefore to understand both the human and product side of the equation. The previous chapter presented a general review of the sensory systems primarily responsible for mediating sensations elicited by foods and beverages. The ways these sensations are quantified comprise the heart of the sensory evaluation side of chemometrics which, in turn, is a specific application of psychophysics.

Two fundamental properties of human beings as sensors are their ability to detect stimuli (i.e., physical energy) and changes in stimuli. This chapter will describe and discuss ways in which the so-called absolute and differential sensitivity can be quantified. Psychophysicists will realize this involves the classical psychophysical methods, and sensory evaluation practitioners will recognize this material as involving paired-comparison, triangle, duo-trio, etc. tests. While the names we call these methods vary, the underlying principles do not. This chapter will also use linear regression to describe results. Regression is discussed in detail in Chapter 9, and the reader may wish to review that chapter now, but that is not necessary to follow the present chapter. For now it need only be said that a regression line is a mathematically determined line which comes as close as possible to each of a set of data points through which it is drawn. The value R^2 is a measure of how close the line comes to the points and varies from 0.0 to 1.0, with 1.0 meaning all points are exactly on the line.

II. THRESHOLDS

The *absolute threshold* is defined as the smallest amount of stimulus energy required to produce a sensation. *Differential thresholds* are defined as the smallest change in stimulus energy that can be detected. This is often referred to as the *just noticeable difference* or *jnd*. It should be noted that thresholds are defined in terms of the stimulus. However, they are quantified in terms of proportion of subjects who (or proportion of occasions one subject) can detect the presence of a stimulus or a change in it, that is, thresholds are statistical concepts.

This is necessitated by the fact that no two human observers are alike, and a given observer varies in sensitivity from instant to instant. The problem of determining thresholds is therefore a problem of handling variability among and within human observers after the data are collected. A practical consequence of this variability is that thresholds must be considered only as approximate: they will be found to vary somewhat from one determination to another.

III. COLLECTING THRESHOLD DATA

Numerous techniques are available for determining thresholds. These have been extensively reviewed in both psychophysical literature[1-5] and in specific sensory evaluation applications.[6-9] The most frequently used methods involve some sort of direct comparison between two or more samples. This can be a comparison between a sample containing the substance of interest and one not containing it or between samples having nonzero but different levels of the substance. For example, it may be of interest to specify a quality control limit for variation in thickener added to a beverage product. This can be done in terms of how likely a given variation in thickener will be noticed as an increase or decrease in perceived viscosity. The problem involves identifying the proportion of consumers who can notice viscosity changes as thickener varies from the intended level. The specific designs of tests developed to approach this sort of problem have recently been discussed in detail by Civille.[7] The concentration here will be on the analysis and interpretation of the numbers taken from such a test. It will be seen that the analysis and interpretation can be extended to data collected by most comparison techniques.

A. INFORMATION DERIVED FROM THRESHOLD MEASUREMENTS
While the main point of interest may be determining a threshold, several other useful pieces of information can be extracted from information collected for that purpose. This is especially true in a multivariate system such as even a simple food or beverage. In fact, the same procedures used to determine a threshold may be used when some of this other information is the major point of interest. The additional information includes information on how variation along one physical dimension will affect the perception of other product attributes (e.g., how will total flavor change as only sugar concentration is changed?), simple sensory-physical and sensory-sensory functions, and an indication of error or bias in the data. This wealth of data is a benefit derived from using multivariate systems and makes up for the relative difficulty that working with such systems can sometimes involve.

IV. THE THRESHOLD STUDY: PAIRED COMPARISONS IN THE METHOD OF CONSTANT STIMULI

The method of constant stimuli is a classical technique for determining both absolute and difference thresholds. While the technique is ideally suited to the use of pair testing, it can be used with any of the difference testing methods commonly employed.[7] However, the practitioner will need to decide if the added complexity of some of the difference testing methods warrants their use.[10]

The observer's task is to decide which of two samples presented has more of a given attribute or attributes. Typically, one standard sample is paired with each of several comparison samples. Often, a total of five to nine different standard vs. comparison pairs are tested. The method can be used to determine a threshold for one given observer, but its true value lies in its ability to quickly collect a large volume of data from a large group of observers. This allows a great deal of generalization to the population because the results are the average of a large group's responses. When large groups are tested, it is not necessary for each observer to taste each pair of samples. For example, if nine pairs of samples are used, each observer need taste only a subset (say, three pairs) and thus avoid sensory adaptation or cognitive fatigue (boredom). It is only necessary that the sample pairs in the subset be randomly assigned and distributed across the observers such that each pair is tasted about equally often across the entire study.

In the example used here, the interest was in determining how several attributes of a simple beverage could be expected to change as attempts to adjust sugar concentration are made, that is, difference thresholds are being determined. The attributes of interest were orally perceived thickness, sweetness, sourness, tartness, and orange flavor. The standard beverage formula was 0.5% orange flavor (MCP Orange Durarome, #3419), 0.3 M (10.27%) sucrose, 0.0078 M (0.15%) citric acid, and gum arabic to 3 centipoise (cps) viscosity. The beverage was made up in tap water. The study was designed to determine how oral perception of this product would change as sugar concentration varied around the standard concentration. This is a simple multivariate problem that can be attacked using basic paired comparison methodology.

Eight comparison samples were prepared. These were identical to the standard in every way except for the concentration of sugar which ranged from a low of 0.12 M (4.1%) to a high of 0.48 M (16.43%). The range was determined by adding 0.045 M (1.54 %) sucrose to each successive sample starting with the lowest concentration. Note that the standard formulation is also used as a comparison. The increment of 0.045 M sugar concentration was chosen because differences in the sweetness of sucrose are detectable with concentration changes of around 10 to 20%.[10] In a complex system, it can be expected that the detectable change will tend toward the high end of this range. It is important to ensure that the lowest and highest concentrations of the material of interest are almost always judged different from the standard and that samples nearer to the standard are increasingly difficult to discriminate from it. This ensures that a smooth sensory-physical function will emerge from the data and allow reliable interpretation of the results. In practice, a very small pilot testing of the samples and threshold values obtained from the literature[11] will make sample selection quite easy.

A total of 150 observers was recruited. Each observer was given a subset of three pairs of samples. In each pair, one of the samples was the standard and one was the comparison. The pairs were given to observers so that each one was judged about equally often over the course of the study. They were also arranged such that, for each pair, the standard was tasted first on half of the total tastings, and the comparison was tasted first on the other half. The instructions and response sheet given to observers are shown in Table 1. In this study, the observers were told to taste the samples only once in the order indicated. That procedure loosely approximates a real world situation in which the taste of one product may be compared to the *memory* of another tasted hours, days, or longer ago. However, enhanced sensitivity can be built into the data by allowing careful tasting and retasting with thorough mouth rinsing between samples. This is especially true if trained, expert tasters are used. In that case, the increased sensitivity may allow for use of fewer observers. The cost of the increased sensitivity is the loss of generalizability to the general population because of the decreased number of observers and heightened discriminability of the procedure.

A. HANDLING THE DATA

The raw data from this type of study consists of judgments of which sample had more of each of the attributes of interest. These are converted to proportions of total number of judgments made on each pair. Note that this total is less than the total number of all judgments summed across all pairs. This is the case because each of the 150 subjects judged a subset of 3 of the 9 possible standard vs. comparison pairs. Consequently there should be 50 judgments per pair. In the present case, subjects were randomly assigned to the sets of 3 pairs until 150 subjects had completed the study. Slight variation in the number of times each pair was assigned resulted in judgments ranging from 47 to 52 across the 9 possible pairs.

Table 1
TASTE INTENSITY TEST

NAME_____   DATE_____

Today you will be given three different pairs of beverage samples to evaluate. All the beverages will be the same flavor. Each pair will take 2—3 min to evaluate and they will be given to you consecutively as you are ready. When you have finished evaluating one pair, I will give you the next one and you may start it immediately.

In each pair we want to know which beverage has the highest intensity of each of the attributes listed below. Therefore, you must taste both samples before making any judgments.

BEFORE tasting anything, read through the list of attributes so you will know what you are looking for. You only have enough of each beverage to taste it once. We don't want you to retaste.

Please rinse your mouth with the water provided before tasting each sample. Taste the samples in the order given and taste both samples before making any choices. Then, place an "X" in the column below the appropriate beverage to indicate which has the highest level of each attribute. You should only have a response in one column or the other for each attribute.

Tasting Order_____

	Beverage_____	Beverage_____
Which beverage is		
Most sweet?	_____	_____
Most sour?	_____	_____
Most tart?	_____	_____
Thickest? (as felt in your mouth)	_____	_____
Which has the most orange flavor?	_____	_____

Thank you.

The proportion of time each comparison was judged to have more of each attribute than the standard is shown in Table 2. These data should be inspected for internal consistency before proceeding. First, since equal numbers of comparison stimuli had more and less sucrose than the standard, the comparison should be judged more sweet than the standard on an average of about 50% of the trials. Table 2 shows that this is the case. Second, knowledge of mixture suppression[12] leads to the prediction that the sourness of the constant concentration of citric acid used here should decrease as sucrose concentration increases. This should result in fewer judgments of "more sour" as the comparison sucrose concentration increases. This is also the case. Finally, although not evident from the table, on those sample presentations when the standard was tasted first, the standard was judged "greater" on 49.98% of the trials; on those presentations when the standard was tasted second, it was judged "greater" 52.49% of the time. Thus, very little, if any, systematic order bias entered the data, that is, judgments did not depend on which sample was tasted first. In fact, all attributes judged were judged greater than the standard on roughly 50% of all trials, suggesting systematic relations between each one and sucrose concentration. In general, the data conform to known relationships among sensations and are internally consistent. They can therefore be presumed to provide a good database for reliable conclusions and predictions.

B. THRESHOLD DETERMINATION

Figure 1 is a plot of the sweetness data from Table 2. The figure shows the proportion of

TABLE 2
Proportion of Times Each Comparison Sample was Judged to Have More of Each of the Attributes Listed Than the Standard

Sucrose concentration of comparison (M)	Attributes					Number of judges
	Sweet	Sour	Tart	Thick	Orange	
0.12	0.02	0.86	0.60	0.21	0.06	52
0.165	0.02	0.76	0.60	0.12	0.12	50
0.21	0.12	0.69	0.60	0.30	0.23	52
0.255	0.27	0.67	0.73	0.35	0.55	51
0.30 (std.)	0.40	0.56	0.56	0.50	0.40	52
0.345	0.71	0.41	0.45	0.69	0.49	49
0.39	0.85	0.29	0.44	0.73	0.65	48
0.43	0.90	0.20	0.31	0.78	0.71	49
0.48	0.89	0.21	0.36	0.77	0.66	47

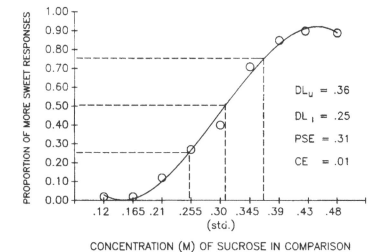

FIGURE 1. Difference threshold data for sweetness of an orange beverage. The difference thresholds were determined by estimating the sugar concentrations on the X-axis corresponding to the 75 and 25% points on the Y-axis.

occasions on which the comparison was judged more sweet than the standard. The result is an example of the classic S-shaped curve, or *ogive*, that is obtained with the method of constant stimuli. In this instance, the curve was fitted and drawn by computer, but it is easy to see that fitting the curve by hand would be a simple matter with such smooth data. In fact, the regularity of data points seen here is very common with this method.

The simplest and most straightforward way to determine a difference threshold is done by hand. First a horizontal line from the 50% point on the y-axis is extended to the curve. A vertical line is then dropped to the x-axis from the point at which the horizontal line intersects the curve. The point at which the vertical line intersects in the x-axis corresponds to that concentration of sucrose which would be judged more sweet than the standard (as well as less sweet) on 50% of the trials and is called the point of subjective equality, or PSE. The PSE is about 0.31 M sucrose. In general, the PSE will not be exactly equal to the standard due to extraneous influences on the data which introduce a constant error (CE). In this study, the error shifted the PSE to slightly higher than the standard. Some common forms of error will be discussed later. For now it can be seen that the error here is quite small.

Since thresholds are statistical concepts, they can be defined somewhat at our pleasure. The 50% points just plotted to obtain the PSE represents the stimulus concentration that results in no discrimination from the standard, i.e., the subjects are guessing. Conversely, that point at which 100% of the subjects say the comparison is greater than the standard represent perfect discrimination. One reasonable definition of the threshold is that point halfway between pure guessing (50%) and perfect discrimination (100%). Thus, the stimulus which is judged greater than the standard on 75% of the trials is one popular point for threshold definition. Starting at the 75% point on the y-axis, a horizontal line is drawn to intersect the curve, and from there dropped vertically to intersect the x-axis. The concentration of sucrose on the x-axis at the point this line intersects it is that concentration judged greater than the PSE concentration on 75% of the trials. Here, that concentration is estimated at 0.36 *M*. The range between the PSE and this concentration is the upper difference threshold, or difference limen, abbreviated by DL_u. In this case the DL_u is 0.36 *M* – 0.31 *M* = 0.05 *M* sucrose. The CE in this study is only about 20% of the difference threshold. This DL_u is referred to as the upper limen because it designates how much the sucrose concentration must be *increased* before 75% of the subjects will notice the change in sweetness. The same procedure is used to estimate the lower difference limen, or DL_l, that is, a horizontal line from the 25% point on the y-axis is extended to the curve and then dropped vertically to the x-axis. The sucrose concentration at that point corresponds to the concentration which is judged less sweet than the PSE 75% of the time. Here it corresponds to 0.255 *M* sucrose or a DL_l of 0.055 *M* sucrose. The upper and lower thresholds are virtually equal in this study but this is not always the case as asymmetry can result. This is especially true when stimuli close to absolute threshold are studied as the DL tends to increase at very low sensation intensities.[1,13] Also, since thresholds are statistical, it would have been just as reasonable to calculate them with any other degree of difference in mind. For example, we could have based the upper and lower thresholds on 60 and 40% "greater than" values had we chosen to do so.

Since it is common to obtain a very smooth ogive when the method of constant stimuli is used, it is usually sufficient to estimate thresholds by visual inspection and hand-fitted curves. If more precision is desired for any reason, the difference threshold can be calculated mathematically by using z-scores. Whenever the data plot is an ogive, it indicates that differential sensitivity is normally distributed about the standard. In those cases, the proportion of "greater than" judgments can be transformed directly to z-scores and plotted as a function of the comparison stimuli. z-scores are tabulated in any introductory statistics textbook, and transformation is a simple matter of finding the tabled z-score that corresponds to the obtained proportions. This was done and the z-scores are plotted as a function of sucrose concentration in the comparison stimuli in Figure 2. When the points along an ogive are transformed to z-scores, a straight line results. Although Figure 2 still shows some curvature at each end of the plot, a least squares regression line was fit to the data and the R^2 value of 0.97 indicates the data are very well described by a straight line having the formula

$$Y = 10.7X - 3.45 \tag{1}$$

To calculate the PSE, enter the value of Y in the equation which corresponds to 50% "greater than" judgments. The z-score for 50% is 0. Substituting 0 in the equation and solving for X,

$$0 = 10.7X - 3.45 \tag{2}$$

$$3.45 = 10.7X \tag{3}$$

$$0.32 = X \tag{4}$$

The calculated PSE is 0.32 *M* sucrose. Similarly, the DL_u is found by substituting the z-scores corresponding to a "greater than" value of 75% in the equation and solving for X. The

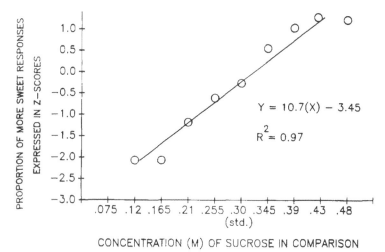

FIGURE 2. Data from Figure 1 converted to z-scores. A straight line coming as near as possible to all data points was then drawn through the points, and the formula for the line can now be used to precisely estimate the difference thresholds (see text).

z-score for 75% is approximately 0.68. Therefore:

$$0.68 = 10.7X - 3.45 \tag{5}$$

$$4.13 = 10.7X \tag{6}$$

$$0.38 = X \tag{7}$$

By the z-score method, the concentration of sucrose corresponding to a DL_u based on 75% "greater than" judgments is 0.38 M. This is obviously very close to that obtained via visual estimation and therefore validates the precision of that technique.

C. WEBER FRACTION

Earlier in this chapter, as well as in Chapter 5, it was noted that a change in stimulus concentration of about 10 to 20% is, as a general rule, needed before a change in taste intensity can be noticed. This value is the *Weber Fraction* for the sense of taste. The Weber Fraction describes the fact, discovered by E. H. Weber (1834),[1] that the change in stimulus intensity needed to produce a just noticeable change in sensation intensity is a constant fraction of the initial stimulus intensity. Mathematically,

$$\Delta I/I = C \tag{8}$$

where ΔI is the change in intensity needed to result in a jnd, I is the initial intensity, and C is the resulting constant. For example, suppose that ten candles are lit in an otherwise unlit room, and the addition of one more lighted candle results in a jnd in brightness. According to the Weber Fraction, if 100 candles are lit in the room, then 10 more would be needed to result in a jnd in brightness, that is, the Weber Fraction is 0.10 or 10%. This fraction serves as a useful tool for prediction of whether formulation changes will result in detectable differences in the sensations likely to be affected. In the example of the orange beverage used here, an increase in sucrose concentration of about 16% would be expected to be noticed by about 75% of the population, that is, the PSE was found to be 0.31 M sucrose and 75% "greater than"

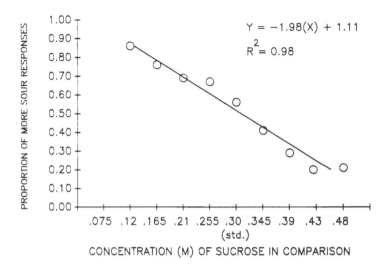

FIGURE 3. Proportion of "more sour" responses in an orange beverage as a function of increasing sugar concentration in the beverage.

judgments corresponded to a sucrose concentration of 0.36 M. Therefore, in the Weber Fraction, $I = 0.31\ M$, and $\Delta I = 0.36\ M - 0.31\ M = 0.05\ M$:

$$\Delta I/I = C \tag{9}$$

$$0.05/0.31 = 0.16 \tag{10}$$

Consequently, in this or a similar beverage system, it can be expected that a 16% increase in sucrose concentration will result in a jnd based on 75% "greater than" judgments. Establishing this relationship avoids the extra effort involved in testing all formulation changes for differences relative to the original standard. Instead, only those that fall beyond the Weber Fraction might be judged to require large-scale testing while smaller changes might require only smaller-scale confirmatory studies.

The Weber Fraction is constant across most of the sensory range of a given sensory system. However, it increases sharply as the sensation in question nears its absolute threshold. Thus, a Weber Fraction determined for a sensation that is not near threshold is likely to underestimate the change needed for a jnd near absolute threshold. The fraction will therefore be conservative near threshold and might even be exceeded without noticeable changes in how the product is perceived.

D. ADDITIONAL INFORMATION
1. Sourness
Given the complex nature of sensory interactions, it is clear that attributes besides sweetness will change as sugar concentration varies. Figure 3 shows how perceived sourness changed in the orange beverage as sucrose concentration varied. Mixture suppression predicts that a constant level of citric acid, as used here, should taste progressively less sour as sucrose concentration and sweetness increase. It is therefore expected that the number of "greater than" sour judgments for the comparison sample will decrease as its sucrose concentration increases. Figure 3 shows this relationship and the straight line that describes it ($Y = -1.98X + 1.11$). The proportion of subjects who reported that the comparison was more sour than the standard decreased as a linear function of sucrose concentration in the comparison. The linear decrease in "greater than" judgments (as opposed to an ogive function as seen with sweetness)

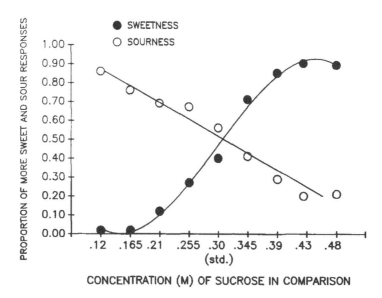

FIGURE 4. Sweet and sour threshold data plotted on the same graph to indicate the inverse relationship between the two.

is due to the fact that citric acid concentration remained constant while sucrose concentration varied.

The same parameters obtained for sweetness can now be obtained for sourness. By substituting the appropriate Y value in the equation for sourness judgments, the PSE is found to be 0.30 M sucrose. that is, the sourness of the standard was judged equal to itself: there was no constant error in sourness judgments. The point at which 75% "greater than" judgments occurred for sourness corresponded to a sucrose concentration of 0.18 M, to result in a DL_u for sourness of 0.12 M sucrose. This means that a *decrease* in sucrose of about 0.12 M will result in about 75% of the population detecting an *increase* in sourness. Similarly, the DL_l is found to be 0.13 M sucrose: *increasing* sucrose by that molarity will result in 75% of the population detecting a *decrease* in sourness.

Figure 4 shows both the sweetness ogive and the sourness linear function on the same plot. This plot, along with the mathematical relationships developed earlier, provides a useful description of how the perception of sweet and sour will change as sucrose concentration varies. The picture is an indirect one, however, because the actual perceived intensity of sweetness and sourness were not measured. So, for example, the plot cannot tell at what point the sweetness and sourness of the beverages are equal. It can only tell us how many people detect changes in these attributes as sucrose concentration varies. However, the judgments depicted in the plot do provide the critical information needed to predict sensitivity to formulation changes.

2. Orange Flavor

The chemical senses literature leads to the expectation that orange flavor might also change with sucrose concentration.[14-16] If this is the case, then the proportion of "greater than" judgments for orange flavor in the comparison sample should also change. Figure 5 shows this is the case. Here the relationship is described by a quadratic equation

$$Y = -0.39 + 4.02X - 3.64X^2 \tag{11}$$

such that judgments of more orange flavor are a slightly negatively accelerated function of sucrose concentration when the orange flavor ingredient is held constant ($R^2 = 0.94$). The

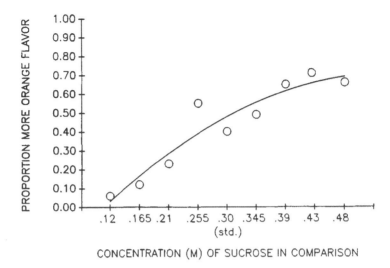

FIGURE 5. Increases in the proportion of "more orange flavor" responses as a function of increasing sugar concentration.

perception of orange flavor is therefore influenced by sweetness, and this knowledge can be used to predict responses to samples being formulated. As sucrose concentration increases, its effect on orange flavor judgments becomes progressively smaller as indicated by the negatively accelerated function, and eventually it masks the orange flavor. By substituting the highest level of sucrose (0.48 *M)* into the quadratic equation as the value of X and solving for Y, it is found that about 69% of the respondents would be expected to report a noticeable increase in orange flavor resulting from only the increased level of sucrose. If threshold is based on 75% "greater than" judgments of orange flavor, this would be defined as below threshold. However, the cut off for threshold is arbitrary, and the relationship can also be stated as indicating a certain proportion of the population is likely to respond to increases in sucrose by reporting an increase in orange flavor. Whether or not the proportion responding in that fashion is meaningful depends upon the judgment of the formulator. As the sweetener level increases and becomes perceptually more distinct in its own right, it will add successively less to the perceived intensity of orange flavor. However, exactly how much will be added to orange flavor cannot be determined from these data. Only *increases* or *decreases* in intensity for the individual attributes can be inferred.

3. Tartness and Thickness

These attributes, like sourness, were not directly manipulated but can be expected to change as a function of sucrose. Tartness can be thought of as a combination of sweet and sour, and perceived thickness is known to be influenced by sweetness.[17] These relationships are both shown in Figure 6. Tartness is described by a quadratic function that curves downward (Y = $0.53 + 1.09X - 3.33X^2$) as sucrose concentration increases. Indirect inference from the proportion of "greater than" judgments suggests that tartness remains constant at all sweetness levels below the standard, but decreases at higher sweetness levels. Recall that low sweetness levels are associated with higher sourness and that sourness decreases at higher sweetness levels. It can therefore be inferred that tartness is influenced more by sourness, while sweetness will tend to decrease tartness. The ratio of sweet:sour therefore seems critical to this attribute, and how sweet and sour are integrated to result in tartness will be discussed later. Here, only the direction of the relationship can be determined because no direct estimates of tartness, sweet, or sour were obtained. Consequently, it can be said from these data that samples with less sucrose were more tart, but how much more tart is not known.

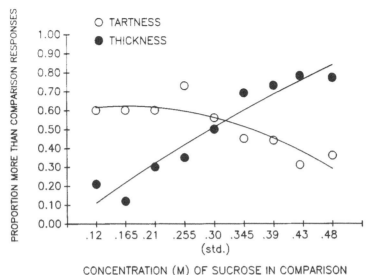

FIGURE 6. Changes in the proportion of "more tart" and "more thick" responses as sugar concentration changes.

Perceived thickness increased as a quadratic function of sucrose ($Y = -0.19 + 2.71X - 1.16X^2$; $R^2 = 0.96$), but could not be accounted for by variation in viscosity which ranged only from 2.3 to 3.2 cps. The very slight curvature of the thickness function indicates that a straight line can also describe these data very well, but a quadratic function was chosen because it was judged that the admittedly small increase in precision was not outweighed by the equally small increase in complexity of the quadratic equation over the linear equation. Sweetness has a strong effect on expected thickness,[17] and this expectation translates into the enhanced perception. This is an excellent example of the fact that sensory systems are active and bring information to the situation, rather than acting simply as passive receivers of external energy.

4. Sensory-Sensory Relations

Sensory-sensory relations describe how one sensation changes as a function of another sensation. This is distinct from sensory-physical (i.e., psychophysical) relations which describe how sensations are related to physical parameters. The former hold a certain degree of generality beyond sensory-physical relations in that they are based on sensations alone. Providing the same sensation intensities and qualities can be achieved with different physical stimuli, the general sensory-sensory relation should hold regardless of the stimuli. For example, if sourness is suppressed to a certain extent by sugar A, it should also be suppressed to the same extent by sugar B when sugar B is adjusted in concentration to match the sweetness of sugar A. Note that this refers to *perceived sweetness*, not to sugar concentration. Two sugars matched in sweetness intensity are very likely to be at different concentrations.[18] It is this sort of relationship that is implicitly spoken of when the sweetening power of various sweeners is compared. To say that one sweetener is 100 times more sweet than another means that it can be perceived at a given intensity (often threshold) at 1/100th of the concentration of the other. The gust scale is another example of sensory-sensory relations among tastes.[19] Sensory-physical relations tend to be much more specific with each compound that produces a given sensation having its own threshold and psychophysical function.[20-22]

Figure 7 shows the sensory-sensory relationships between tartness and both sweetness and sourness plotted in terms of the proportion of "greater than" judgments. Both relationships are well described by quadratic functions ($R^2 = 0.93$ in both cases). Together these plots allow inferences to be made about the contribution of sweetness and sourness to tartness. While the

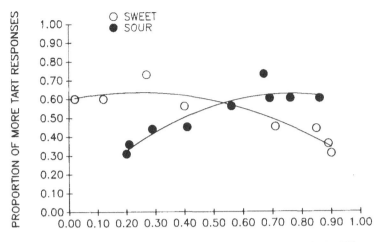

FIGURE 7. Sensory-sensory relations between perceived sweet and sour and perceived tartness. As perceived sweetness (open circles) increases, perceived tartness decreases; as perceived sourness (filled circles) increases, perceived tartness increases.

relative intensities of sweet and sour at a given intensity of tartness cannot be determined from these data, it is clear that comparison samples which were judged to be more sweet than the standard by a high proportion of the subjects also tended to be judged less tart. The reverse is true for sour: increasing the proportion of "more sour" judgments is associated with increasing the proportion of "more tart" judgments. The inference is that sweet and sour are integrated by the taste system in such a way that results in sweet suppressing tartness and sour enhancing it. This same *general form* of relationship should hold, regardless of which compounds are used to produce the sweet and sour sensations.

A convenient property of sensory-sensory relations is that they can be determined in the absence of knowledge of the sample compositions. All that is needed is a series of samples representing a range of sensation intensities and reliable ratings or rankings of those intensities. This property is particularly valuable in more complex systems which may be composed of hundreds of different compounds, many of which give rise to a particular sensory attribute only in combination with others. In these cases it is often much easier to obtain sensory data than it is to chemically characterize the samples. The sensory data derived from sensory-sensory relationships can often serve as guidelines, or at least indicate starting points for chemical analyses.

V. ABSOLUTE THRESHOLDS

The absolute threshold is the lowest level of a stimulus that can be detected. With the method of constant stimuli, this is determined in the same way as is the difference threshold, except that the standard sample contains none of the stimulus of interest. For example, if the detection threshold of the orange flavor used in the example above were of interest, the standard sample would contain no flavor at all and the comparisons would be selected to range in intensity from clearly not detectable to very clearly detectable. Usually detection threshold is defined as that concentration detected 50% of the time. All other procedures are the same as those used in determining difference thresholds.

A distinction is made between detection and recognition thresholds. Very often a compound can be detected at a lower concentration than that needed to produce its characteristic sensation quality. For example, a complex flavor like orange, which has many components, is especially likely to be detected at a concentration below that needed to clearly taste orange. This is because the individual components of the flavor each have their own detection thresholds and it is not until all of them are above threshold that they will be available for the sensory system to integrate into orange flavor. However, even very simple sensations that can be attributed to a single chemical, like sweet, tend to be detected before they are recognized. Therefore, it is customary with the method of constant stimuli to determine both thresholds. Recognition threshold will be somewhat higher than detection threshold.

VI. OTHER METHODS

Other methods are available for determining thresholds and are described in detail elsewhere.[1,2] One of these is called the *method of adjustment* because it requires the subject to adjust the stimulus intensity to a level where it is just barely detectable (absolute threshold) or just noticeably different from a standard (difference threshold). While this method can provide a very precise estimation of threshold when the stimulus is continuously variable, it is very impractical in chemometrics because of the difficulty in preparing such samples. In addition, the method is not suitable for large numbers of subjects because it requires individual testing.

Another method is known as the *method of limits*. With this technique the investigator presents a series of stimuli of increasing intensity to the subject who judges whether or not the stimulus is detectable or has just noticeably increased in intensity. When the subject has made such a judgment, the investigator begins presenting the same series in order of decreasing intensity from the point at which the judgment was made. This is referred as the "up and down" or "staircase" method of limits[23] and allows the investigator to track the threshold over several such reversals or runs. The average value at which the sensation was noticed, or a change reported, is taken as the absolute or difference threshold, respectively. This is well suited to individual subjects and has gained popularity in clinical settings where threshold determinations are used to quantify sensory deficits.[24]

In general, the method of constant stimuli is easiest to adapt to large groups of subjects and can be used to obtain a wealth of data beyond simply the single point of threshold. In addition, it has recently been found useful in determining ideal flavor concentration in preference studies[6] and is well suited to consumer research where large numbers of subjects are available, but each only for a short period of time.

A mechanical issue in handling data collected by either the method of limits or the method of constant stimuli is how to handle judgments of equal intensity or judgments of "unsure", whether a stimulus was different or present. In determining difference thresholds using the method of constant stimuli, it is recommended to not allow judgments of unsure because of this dilemma. For example, if unsure judgments are allowed, plotting an ogive becomes problematic. Should the unsure judgments be equally distributed between the "less than" and "greater than" standard halves of the curve? If they are, it would have been easier to just force a judgment in the first place and not depend on the assumption that the unsure judgments would equally distribute in that fashion. Eliminating the unsure judgments from the database after they have been allowed will decrease base size for judgments nearest the standard, because that is where uncertainty will be greatest. However, this is also where the increased precision of a large database size is most needed. One could choose not to plot an ogive, and instead tabulate for each comparison stimulus the proportion of definite judgments and unsure judgments. The range between the standard and the highest stimulus where unsure judgments last occurred will specify the DL_u, and the lowest stimulus at which the last unsure judgment occurred specifies the DL_l. The interval between these two stimuli is referred to as the interval of uncertainty, or IU. Unfortunately, the IU will vary from subject to subject and therefore no clear cutoff point for it will occur. It must therefore be handled probabilistically, just as the threshold itself is. Additional rationale for not allowing unsure judgments is presented in conjunction with discussion of signal detection theory below.

VII. THE CONCEPT OF THRESHOLD

The concept of a sensory threshold has been discussed extensively.[1-4,25] The threshold is no longer viewed as an absolute point below which no sensation is perceived and above which a sensation is always perceived. It is now recognized that the threshold is a probabilistic concept, must be dealt with statistically, and varies both between observers and within a given observer across time. In addition, threshold values are not solely a property of the sensory system. They are strongly influenced by the expectations brought to the situation by the subject.[25] However, it is also clear that as stimulus energy decreases, the probability of detecting it also decreases, and the degree of that decrease is a useful piece of data. Further, on an intuitive level, it does seem reasonable to believe that there is some energy level that cannot be discriminated from background noise by the observer and at which the observer's responses are almost entirely governed by expectations. Conversely, it is also reasonable to believe that there are energy levels at which the honest observer will always report a sensation. The area between these extremes requires the separation of expectation from sensitivity.

VIII. THE THEORY OF SIGNAL DETECTION

Subjects in threshold studies bring a set of expectations and response tendencies to the task which, to a large extent, govern their behavior. These expectations and tendencies must be separated from the intrinsic sensitivity of the sensory system in order to obtain a true representation of the detectability of the particular stimulus. The Theory of Signal Detection, or TSD, provides a means of separating sensitivity from response tendencies of subjects.[25] The procedures of TSD have their greatest usefulness in detection tasks in which one sample at a time is presented to observers who must decide if the sample contains, or does not contain, the attribute of interest. In procedures involving forced choice paired comparisons, such as the method of constant stimuli discussed above, the proportion of trials on which the subjects correctly identified the sample containing the attribute of interest, or a greater amount of it, can be used as a direct measure of sensitivity. The reasons for this will be described here. The calculations involved in TSD are very straightforward and can be found in numerous sources.[1-3,26] Since the proportion of correct responses in a paired test is a direct measure of sensitivity and is immediately available without further calculation, TSD calculations will not be discussed here in any detail.

Consider the following situation. An observer is presented with a series of samples, one at a time, and for each sample must decide if it contains the attribute of interest before the next sample is presented. The samples are complex, such as foods or beverages, and the attribute of interest (i.e., the signal) is relatively weak. Further, only half of the samples contain the attribute, the rest being controls. In this circumstance a liberal subject would have a tendency to respond "yes", meaning he or she detected the signal, on a relatively high proportion of the trials. With a liberal response tendency, the subject is likely to respond "yes" on a fairly high proportion of trials when the sample does in fact contain the signal attribute, that is, the subject will have a high proportion of hits. The high proportion of hits could be construed as indicating the signal is easily detectable by that subject. However, liberal subjects will also have a high proportion of false alarms. A false alarm occurs when the sample does not contain the signal but the subject said "yes" anyway. A high hit rate in conjunction with a high false alarm rate is the distinguishing mark of a liberal response tendency, and indicates that the subject has a low criterion for saying "yes". A conservative subject will have exactly the opposite pattern

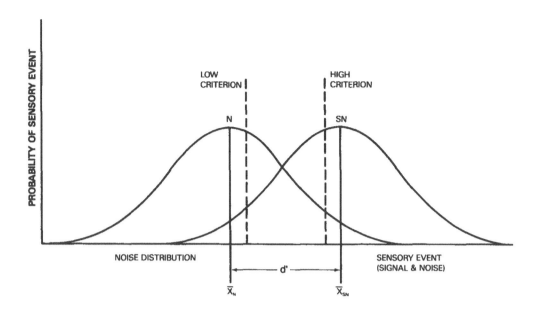

FIGURE 8. Theory of signal detection. The distribution on the left, labeled N, represents sensory noise. The distribution on the right, labeled SN, represents the distribution of sensory events when a signal is imposed on the naturally occurring sensory noise. The difference between the means of these two distributions corresponds to the difficulty of detecting the signal in the presence of noise. As the distributions move farther apart, the signal becomes easier to detect and the index of detectability, d', increases. The vertical dashed lines represent the observer's criterion. Whenever a sensory event is to the right of the criterion, the subject reports a signal. With a high criterion (i.e., one far to the right), very few noise events will be above the criterion, and consequently there will be few false alarms. However, a large number of signals will be below the criterion and judged to be noise. Those are "misses". As the criterion decreases (moves to the left), the number of hits increases, but so does the number of false alarms even though the true detectability of the signal remains constant.

of responding: a low hit rate in conjunction with a low false alarm rate. Such subjects have a tendency to respond "no" on a high proportion of trials. In the terminology of TSD, they have a high response criterion. The TSD explanation of this situation is as follows. Sensory systems are noisy systems which fluctuate in responsivity over time. This noise is especially problematic when weak and/or complex stimuli are being processed by the system. Thus, for every sample presented, the subject must decide if the sensory experience is the result of only a noisy system or if the signal is in fact there. The problem is compounded because the random fluctuations of sensitivity will also affect the perceived strength of the signal such that some true signals will seem like only noise. The subjects cope with this problem by setting a response criterion. Whenever the sensory event exceeds the criterion the subject will respond by saying "yes", meaning a signal was detected. If the sensory event does not exceed the criterion, the subject will say "no". When the criterion is low, a large proportion of trials on which only noise occurred will be responded to with "yes", but a large proportion of trials on which the signal was present will also be responded to with "yes". Conversely, when the subject has a high criterion, fewer noise trials will be responded to positively, but fewer signal trials will also be responded to positively. Figure 8 presents this situation schematically. Assuming that the responsiveness of the sensory system to both signal and noise varies randomly and is normally distributed, the subjects criterion and sensitivity can be described in terms of the normal curve. As the subject's criterion increases, a greater proportion of the signal distribution falls below it and will be judged erroneously to not contain the signal, with the result being fewer hits. As the criterion decreases, a greater proportion of the noise distribution will fall above it and be erroneously judged to actually be a signal resulting in more false alarms. Consequently, the criterion can be determined by comparing hit rates to false alarm rates. The criterion is referred to as β and is obtained by converting the proportion of hits to the corresponding ordinate value on the normal curve and dividing that value by the proportion of false alarms converted to an ordinate value on the normal curve. Sensitivity is referred to as d' in TSD and is obtained by the formula:

$$d' = Z_N - Z_{SN} \qquad (12)$$

where $Z_N = 1.0 - p(\text{false alarms})$ and $Z_{SN} = 1.0 - p(\text{hits})$. This simple subtraction provides an estimate of the difference between the mean of the noise distribution and the mean of the signal distribution expressed in standard deviation units. Since it directly compares the means of the two distributions, it is a measure of sensitivity uncontaminated by the subject's criterion. The criterion, or β, is obtained by:

$$\beta = \text{Ordinate SN/Ordinate N} \qquad (13)$$

where SN is the ordinate value of the normal curve corresponding to the hit rate, and N is the ordinate value of the normal curve corresponding to the obtained false alarm rate. Studying this equation will help in understanding what is meant by a high or low criterion. For example, if the subject has a large number of false alarms and hits, it can be seen from the equation that the criterion will be low. This means the subject is trigger happy and is very willing to say a signal is present. Conversely, a high hit rate in conjunction with a low false alarm rate indicates a high criterion, but great sensitivity to the signal.

The forced choice paired comparison procedure avoids the necessity of these calculations by making a simple and reasonable assumption, that is, it is assumed that the sample which does not contain the attribute of interest represents the noise distribution, and the sample which contains the attribute of interest represents the signal distribution. By tasting and comparing both samples, the subject *directly* compares signal to noise and need not make any judgment which involves his or her personal criterion. Therefore, percent correct judgments in a forced choice paired comparison study is a direct estimate of signal detectability.

It should be noted that allowing the subjects to make unsure judgments, or judgments that neither sample contained the signal in a forced choice procedure, eliminates that procedure's advantages; that is, as soon as unsure judgments are allowed, the subject is placed in a situation where he or she is judging the magnitude of difference between the two samples against his or her criterion for saying that a difference exists. Consequently, unsure judgments allow all the uncertainty of a variable criterion to enter back into the study. This is another argument in favor of not allowing unsure judgments to be given in the method of constant stimuli. The investigator may have a compelling reason to determine the degree of certainty with which subjects in a signal detection experiment are responding. If this is the case, procedures for doing so within the signal detection framework have been described[26-29] and used in sensory evaluation. Although the procedures are relatively straightforward, they are not as simple as a direct forced choice paired comparison, and still leave unresolved the question of what to do with the uncertain judgments other than quantify their occurrence.

The considerations of TSD are of critical importance in selecting subjects for various sensory studies. For example, within a product development project, members of the project team will have various expectations and desires concerning formulations of the product and their responses will be heavily influenced by those expectations and desires. Proponents of a certain formulation will have a low criterion for undesirable flavors in competing formulations and a high criterion for the same flavors in their own formulation. Conservative managers who are responsible for product profitability are also likely to have low criteria for negatives. Such judges will have a high hit rate for negatives, which will go undetected unless their false alarm rate is also determined or a paired comparison procedure is used with forced choice. The same problem in reverse will occur for project leaders who are anxious to see their product reach the market. TSD provides a practical set of guidelines for observer selection in product development work and also provides a way to ensure that sensitivity is being measured without the influence of personal criteria.

REFERENCES

1. **Gescheider, G. A.,** *Psychophysics: Method, Theory, and Application,* 2nd ed., Lawrence Earlbaum Assoc., Hillsdale, NJ, 1985.
2. **Baird, J. C. and Noma, E.,** *Fundamentals of Scaling and Psychophysics,* John Wiley & Sons, New York, 1978.
3. **Kling, J. W. and Riggs, L. A., Eds.,** *Woodworth and Schlosberg's Experimental Psychology,* 3rd ed., Holt, Rinehart, & Winston, New York, 1972.
4. **Corso, J. F.,** A theoretico-historical review of the threshold concept, *Psychol Bull.,* 60, 356, 1963.
5. **Simpson, W. A.,** The method of constant stimuli is efficient, *Percept. Psychophys.,* 44, 433, 1988.
6. **McBride, R. L. and Booth, D. A.,** Using classical psychophysics to determine ideal flavour intensity, *J. Food Technol.,* 21, 775, 1986.
7. **Meilgaard M., Civille, G. V., and Carr, B. T.,** *Sensory Evaluation Techniques.* Vol. 1, CRC Press, Boca Raton, FL, 1987.
8. **Moskowitz, H. R.,** *Product Testing and Sensory Evaluation of Foods.* Food & Nutrition Press, Westport, CT, 1983.
9. **Amerine, M. A., Pangborn, R. M., and Roessler, E. B.,** *Principles of Sensory Evaluation of Food,* Academic Press, New York, 1965.
10. **Pfaffman, C., Bartoshuk, L. M., and Mcburney, D. H.,** Taste psychophysics, in *Handbook of Sensory Physiology,* Vol. 4, Beidler, L. M., Ed., Springer-Verlag, New York, 1971.
11. **Fazzalari, F. A., Ed.,** *Compilation of Odor and Taste Threshold Values,* ASTM, Philadelphia, 1978.
12. **Bartoshuk, L. M.,** Taste mixtures: is mixture suppression related to compression?, *Physiol. Behav.,* 14, 643, 1975.
13. **Engen, T.,** Psychophysics: discrimination and detection, in *Woodworth and Schlosberg's Experimental Psychology,* 3rd ed., Kling, J. W. and Riggs, L. A. Eds., Holt, Rinehart, & Winston, New York, 1972.
14. **Murphy, C. and Cain, W. S.,** Taste and olfaction: independence vs. interaction, *Physiol. Behav.,* 24, 601, 1980.
15. **Frank, R. A. and Byram, J.,** Taste-smell interactions are tastant and odorant dependent, *Chem. Senses,* 13, 445, 1988.
16. **Hornung, D. E. and Enns, M. E.,** The contributions of taste and smell to overall intensity: a model, *Percept. Psychophys.,* 39, 385, 1986.
17. **Christensen, C. M.,** Texture-taste interactions, *Cereal Foods World,* 22, 243, 1977.
18. **Moskowitz, H. R.,** Ratio scales of sugar sweetness, *Percept. Psychophys.,* 7, 315, 1970.
19. **Beebe-Center, J. G.,** Standards for use of the gust scale, *J. Psychol.,* 28, 411, 1949.
20. **Moskowitz, H. R.,** Scales of Intensity of Single and Compound Tastes, dissertation, Harvard University, Cambridge, MA, 1968.
21. **Meiselman, H. L.,** Human taste perception, *Crit. Rev. Food Technol.,* April, 89, 1972.
22. **Engen, T.,** *The Perception of Odors,* Academic Press, New York, 1982.
23. **Cornsweet, T. N.,** The staircase method in psychophysics, *Am. J. Psychol.,* 75, 485, 1962.
24. **Slotnik, B. M., Wittich, A. R., and Henkin, R. I.,** Effect of stimulus volume on taste detection threshold for NaCl, *Chem. Senses,* 13, 345, 1988.
25. **Green, D. M. and Swets, J. A.,** *Signal Detection Theory and Psychophysics,* John Wiley & Sons, New York, 1966.
26. **O'Mahoney, M.,** Salt taste sensitivity: a signal detection approach, *Perception,* 1, 459, 1972.
27. **O'Mahoney, M., Kulp, J., and Wheeler, L.,** Sensory detection of off-flavors in milk incorporating short-cut signal detection measures, *J. Dairy Sci.,* 62, 1857, 1979.
28. **O'Mahoney, M.,** Short-cut signal detection measures for sensory analysis, *J. Food Sci.,* 44, 302, 1979.
29. **O'Mahoney, M., Garske, S., and Klapman, K.,** Rating and ranking procedures for short-cut signal detection multiple difference tests, *J. Food Sci.,* 45, 392, 1980.

Chapter 7

MEASURING SENSATION MAGNITUDE: SCALING METHODS

I. INTRODUCTION

Chapter 6 discussed how judgments from an executionally very simple paired comparison test could be used to make inferences about the intensity of sensory attributes. Such data are very useful in predicting the likelihood that changes in formulation will be noticed. In addition to that information, it is almost always desirable to have an idea of the relative intensity of sensations elicited by a sample product. For example, in addition to learning what percentage of the population is likely to detect changes in orange flavor intensity and sourness when sugar concentration is increased by a certain amount, it is also desirable to know about the sensations themselves. How much more intense is the orange flavor? How much more sweetness is tasted? How much has sourness decreased? Is the product more sweet than it is sour? Is the orange flavor more intense than the sweetness? These are very different questions from those concerning how many observers notice differences. These questions ask about the relative intensities of sensations and answering them requires scaling, or measuring, the sensations themselves. This chapter deals with the practical aspects of directly measuring sensation.

II. MEASUREMENT SCALES

Measurement is the assignment of numbers to objects or events according to rules.[1,2] In the present case, the events of interest are sensations or, more exactly, sensation intensities. The operation of assigning numbers is commonly referred to as scaling or sensory scaling. The result of a sensory scaling procedure is a set of numbers arranged such that the properties of the numbers and the relations among them correspond in a meaningful way to the relations among the sensations being scaled.

A wide variety of techniques for scaling sensations is available. Some of the earliest methods involved measuring successive jnds along a sensory continuum by using differential sensitivity techniques as described in Chapter 6. The resulting scale units represented the number of jnds above threshold a given sensation was. Other techniques involved the assumption that equally often noticed differences are equal; that is, if sample A was judged greater than sample B in orange flavor 75% of the time, and sample C was judged greater in orange flavor than sample B 75% of the time, then the difference in orange flavor between samples A and B was assumed equal to the difference between samples B and C. Still other methods involved sorting samples into categories. For example, an observer would be asked to sort a series of samples into seven or nine categories, such that the samples within each category were equal and the perceptual distances between categories were also equal. The concepts and principles that evolved along with these techniques provide important background material and are detailed by Torgerson in the classi *Theory and Method of Scaling*.[3] This book provides a detailed description of scaling practices up to the late 1950s and is a valuable resource. The methods themselves can be cumbersome for the chemical senses, since they often in-volve the preparation of large numbers of samples and require extensive and repeated tastings by the subjects. As a consequence, work in chemometrics is generally done using the more direct rating scale methods that have become popular since the late 1950s.

TABLE 1
Stevens' Classification of Measurement Scales

Scale	Basic operation	Transformations	Statistics	Examples
Nominal	Determination of equality	One-to-one substitution	Number of cases, mode	Numbering of football players
Ordinal	Determination of greater or less	Any operation preserving order	Median, percentiles	Hardness of minerals, quality of lumber
Interval	Determination of equality of differences	$x' = ax + b$	Mean, standard deviation	°F, °C, energy
Ratio	Determination of equality of ratios	$x' = ax$	Geometric mean coefficient of variation	Length, width

Note: The first column lists the type of scale and the second column lists the operation needed to produce the scale. The operations shown in the third column are those which can be performed on the scale values without changing their properties. Some of the statistics Stevens listed as "permissible" for each type of scale data are in the fourth column, and examples of each scale are in the fifth. x' = transformed scale value, x = original scale value.

From Stevens, S. S., *Handbook of Experimental Psychology*, Stevens, S. S., Ed., John Wiley & Sons, New York, 1951. With permission.

III. CLASSIFYING SCALES OF MEASUREMENT

As noted above, assigning numbers to events or objects according to some set of rules results in a measurement scale of the objects. The rules used to assign the numbers determine the properties of the scale and what it tells about the objects it purports to order. Stevens[1,2] proposed a classification system for measurement scales which has become generally accepted as a useful system for ordering scale properties. However, it is not the only system and Stevens was not the first to discuss the issue.[4] Since other systems are basically similar, Stevens' system will be described and the interested reader may refer to Reference 3 for additional detail.

Stevens' system is presented in Table 1. The table shows the names of the scales, how they are determined, their mathematical properties, permissable transforms, and some examples of each. Nominal scales are the simplest form of scale available. Nominal scaling uses numbers as labels. For example, a set of citrus beverage products may be classified as "mostly orange-like" and "mostly lemon-like" by using a "0" to represent orange flavor and a "1" to represent lemon flavor. Since the numbers used are given no meaning other than the labels applied, any other two numbers can be substituted with no loss of meaning; that is, it could be decided to use "2" for orange and "3" for lemon with no loss of information as long as the transformation from "0" and "1" to "2" and "3" is done consistently.

Ordinal scales begin to provide information concerning quantity. An ordinal scale is a rank ordering of samples from least to greatest amount of an attribute. Ordinal scales provide information on intensity order only. For example, three samples of tea may be rank ordered for astringency in the order $A < B < C$. This means that sample A has the least amount of astringency and sample C the greatest. However, no information is available about how large the differences between the samples are. The difference between A and B could be perceptually quite large, while B and C may be quite close together. Ordinal scales, even if they use successive numbers to indicate ranking (e.g., 1, 2, and 3, instead of A, B, and C), only provide a rank order. Stated simply, the size of the differences among numbers on the ordinal scale does not reflect the sizes of differences among the sensations ordered. The numbers on an ordinal scale may be transformed (i.e., changed) in any fashion that maintains the rank order and no information will be lost.

Interval scales provide information on the magnitude of perceived differences among samples being scaled as well as on their rank ordering. Because of this additional information, interval scales are more useful than ordinal scales. On an interval scale the differences between numbers have meaning. For example, a sample rated "7" is two units larger than a sample rated "5" for a given attribute. Further, the magnitude of difference between the numbers on the scale corresponds to the magnitude of difference between the sensations themselves, such that large numeric differences indicate large perceptual differences and small numeric differences indicate small perceptual differences. The zero point on an interval scale is arbitrary, and equal differences on an interval scale represent equal differences in sensation no matter where on the scale they are located. For example, $7 - 5 = 20 - 18$.

Ratio scales provide a still higher level of measurement. Ratio scales provide information on rank order, equal differences (intervals) between samples, and equal ratios among samples. For example, if two tea samples, A and B, are rated on a ratio scale as "1" and "10", respectively, for astringency, they stand in the same ratio as samples C and D which are rated as "10" and "100". This is true because $1/10 = 10/100$. Stated verbally, sample B is ten times as astringent as sample A, and sample D is ten times as astringent as sample C. In addition, we have the rank ordering of these samples: $A < B = C < D$. Finally, we know that, although the ratio of A/B equals the ratio of C/D, the interval between C and D is much greater than the interval between A and B. Because ratio scales provide ratio, interval, and ordinal information they are very desirable. In order to preserve the ratio property of these scales, the only permissible transformations of the data points are those which preserve the ratios among them. For example, the points "5" and "10" may be multiplied by 2 without destroying their ratio. However, adding 2 to each will destroy their ratio: $5/10 = 10/20$, but $5/10 \neq 7/12$.

Table 1 also provides an indication of some permissible statistics to be used in conjunction with each scale type. On the descriptive level, the listed statistics make intuitive sense. For example, on a ratio scale the numbers 1, 10, and 100 have a geometric mean of 10. Using the geometric mean for a ratio scale makes sense because the geometric mean is the mathematical balancing point on a ratio scale. However, if the numbers 1, 10, and 100 are on an interval scale which does not possess ratio properties, the mathematical balancing point is the arithmetic mean or 37.

On the level of inferential statistics the issue of appropriate procedures has been controversial,[5-8] especially with regard to the use of parametric statistics such as t-tests and the analysis of variance. One side of the debate argues that values used in parametric procedures must be normally distributed, error variations must be equal and random for groups being compared, and the data must be on at least an interval scale. According to this view, t-tests and other parametric procedures are inappropriate for data possessing only ordinal properties. The other viewpoint is that the type of scale on which the values lie is not relevant to the type of analysis performed. The references listed provide the details of this debate which is beyond the scope of this book.

From the present perspective, the type of scale data obtained is more relevant to the sort of conclusions that can be drawn about products. For example, if products are rated only on an ordinal scale, then conclusions based on the size of differences between scale values will become problematic. On the basis of ordinal data, the investigator can only expect that the rank ordering of products rated on the scale will remain the same across different rating sessions; the differences between corresponding scale values is likely to change. Therefore, predictions or conclusions based on magnitude of differences will be shaky. With interval data, the differences among ratings are meaningful, but ratios are not.

IV. SENSORY SCALING IN PRACTICE

The overwhelming majority of sensory data in chemometrics is collected by asking observers to taste or smell a sample product and rate the perceived intensities of various attributes on some sort of rating scale. A variety of rating scales is used and many sources describing them are available.[9,10] The most commonly used rating techniques are the various forms of category scales and magnitude estimation procedures. Category scales most often require subjects to assign a single number from a fixed range of numbers to the sensation intensities experienced. The range of numbers typically depends on the individual worker's judgment, but is often around nine categories. However, 20 or more categories can be, and frequently are, used. Magnitude estimation requires subjects to assign numbers to sensations in proportion to their intensity. For example, if one sensation is rated "10" and the next one is three times as intense, it is rated as "30". No limit to the size of numbers is given in magnitude estimation.

Considerable academic debate has taken place regarding which of these general procedures is the correct approach. The debate revolves around what the properties of the underlying perceptual scale are likely to be. Proponents of category scales tend to believe that humans inherently use interval scales and judge differences among sensations,[11] while proponents of magnitude estimation believe that humans inherently judge ratios among sensations.[12] This is a fascinating debate, and consulting the listed references will help the reader decide under which conditions each approach should be used.

While various experiments might suggest that humans use either interval or ratio scales when rating sensations, there is no *a priori* reason why they would always use only one scale type to the exclusion of all others. In fact, there is no *a priori* reason why properly instructed subjects could not use either an ordinal, interval, or ratio scale as desired by the experimenter. It is the present position that useful data can be obtained in practical situations from both category ratings and magnitude estimation with relative ease when the procedures are properly employed.

What about data suggesting otherwise?[13,14] Finding an exception to a rule does not invalidate the rule. For example, in chemistry, Dalton's Law of Partial Pressure states that the total pressure of a mixture of gases in a container is equal to the sum of the pressures of each of the individual gases; that is, each gas exerts a pressure as if it were the only one in the container. If it were found that two gases interacted with each other such that their total pressure did not equal the sum of the individual pressures, Dalton's Law would not be invalidated. Rather, a limiting condition would be stated, i.e., the law holds when the gases in the mixture do not interact. Likewise, if an experimenter finds that category ratings or magnitude estimates vary with certain conditions, the scales themselves are not invalidated. Instead, such results define the limiting conditions which determine the properties of the resulting scale values. These conditions need to be taken into account when interpreting the data. The limiting conditions reflect the operating characteristics of the human measurement device and cannot be ignored any more than can the operating characteristics of any physical instrument.

Several variables which affect the results of category rating and magnitude estimation studies have been identified and described.[15-19] Ways to account for these variables have been proposed and will be discussed here. While these variables are sometimes referred to as biases,[15] the term preferred here is *limiting conditions*. The term *bias* implies that something is wrong with the data, and they should be held in suspicion if not rejected outright. *Limiting conditions* is more neutral and implies the results are being interpreted within the constraints of the experimental situation. This chapter discusses the use of category ratings and magnitude estimates in light of some of the stronger limiting conditions affecting these procedures. More detailed discussions can be found in several sources.[11,12,20]

A. CATEGORY RATINGS

Category rating here means the assigning of a number from a fixed and finite range of numbers to represent the intensity of a given sensation. Very commonly, category scales consist of the numbers "1" through "9", or a graphically depicted scale consisting of 9 consecutive boxes corresponding to increasingly greater sensation intensities. However, verbal labels describing increasing intensity levels from, for example, "very slight" to "very great" intensity are also common. After data collection, the verbal labels are converted to numbers for analysis. Civille[9] describes the numerous forms of these scales in detail and should be consulted as a resource.

Rating scale data is often collected and used at face value as if the ratings reflect true and unvarying differences among products and as if the differences between categories are equal, that is, they are assumed to have equal interval properties. It is not uncommon for repeated use of category ratings to lead to the realization that product ratings and differences between products change across experiments. This situation may lead one to reject rating scales as unreliable. In some cases, the researcher may conclude that only personal judgment is available to predict consumer responses to products.

In fact, the often observed variability in category ratings follows simple and well-documented rules. An understanding of these rules will simplify things greatly. Category ratings are strongly influenced by the context established by the samples being rated, and about 80% of these contextual effects can be accounted for by the range and frequency of the samples.[19,21]

The *range* of samples refers to the difference between the two extremes, i.e., the lowest and highest perceived intensity samples. When using category rating scales, humans have a strong tendency to match the range of samples to the range of available categories. For example, if a given set of five samples are rated on a nine-category scale, the highest sample will receive a rating somewhere around "9" and the lowest sample will be rated somewhere around "1". If the *sample* range is now changed such that the previous highest sample is now in the middle of the range and two even more intense samples are rated by the same or different observers, then the new highest sample will be rated around "9" and *all* the other samples will receive lower ratings, including the previously highest sample. Figure 1 shows this effect.[22] Note in the figure that the number of samples rated is constant and only the range of samples has changed. The shorter ranges result in steeper sensory-physical (i.e., psychophysical) functions, and the ratings of samples common to more than one range actually change. Changing the number of categories available for rating will have a similar effect.

The *frequency* of samples refers to the *number* and *spacing* of samples. For example, a given intensity of sensation may be represented a greater number of times than other intensities, or a given subrange of the intensity continuum may have more samples occurring within it than other subranges, even though each individual intensity is represented only once. Both of these situations will affect category ratings because humans exhibit a strong tendency to use all of the available categories about equally often. The effects of frequency variation are very evident when category ratings are plotted as a function of the physical variable of interest. In such plots, the areas of the sensory continuum which correspond to the areas of physical continuum having more samples, or more closely spaced samples, will exhibit a steeper slope.[19,21]

Under good conditions, range and frequency of samples can be controlled by the researcher. For example, when flavor intensities/concentrations are being evaluated for effect on total formulation, the actual range and frequency of prepared samples can be predetermined. On the other hand, if a series of samples from the open market is being evaluated, then the sensory evaluation worker may have only an incomplete and superficial idea of the range and frequency of available samples and their attributes. Whether range and frequency are controlled or not, they will affect the ratings. From a practical standpoint, the effect is evidenced by the attribute ratings of a given product changing whenever it is evaluated in the context of

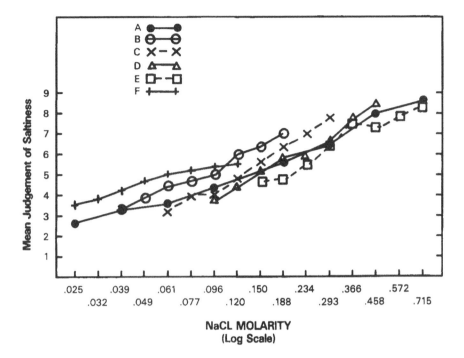

FIGURE 1. Effects of sample range on category ratings. The Y-axis represents category ratings of perceived saltiness and the X-axis represents concentration of NaCl in each sample. Note that the same NaCl concentration will receive different saltiness ratings depending upon the range of the other samples it is tasted with. A narrower range of sample concentrations results in a higher rating for a given sample, and a sample contained in a narrow range at the low end of the overall range is also rated higher than when it is contained in a range extending to the higher end of the overall range. (From Riskey, D. R., *J. Sensory Studies*, 1, 217, 1986. With permission.)

one or more new samples. Including one standard sample that is common to all sets of samples rated will only partially solve the problem because the ratings of the standard are also subject to range/frequency effects. That sample ratings and differences between sample ratings fluctuate in this manner is evidence that category ratings taken at face value only provide ordinal data. Recall that this means only the rank ordering of samples on the scale is meaningful and differences between samples are not.

In many applications, an ordinal rating of the samples is sufficient. In other cases, a meaningful measure of differences between samples that is independent of range and frequency of samples is desired. This is the case, for example, if progress toward a formulation goal must be documented.

A simple and effective way of determining values that are independent of range and frequency of samples is available.[19] The model for doing this is called the *range-frequency model*. According to this model, the true rating of a sample that would be obtained if no frequency effects were present is called the *range* value for that sample and is symbolized by R_i. It is assumed that the R_i depends solely on the relationship between the sample it corresponds to and the two endpoint samples. This value will be relatively uninfluenced by changes in sample context and is therefore desirable to obtain. The *frequency* value for a given sample, called F_i, is the value the sample would have been given if the samples were maintained in proper rank order and each available category is used equally often. Even though all categories are not, in fact, always used equally often,[15] especially the extremes, the deviation is not great enough to invalidate the model, and simple techniques to more nearly equalized category use are available.[15]

Given the above assumptions and knowing beforehand the number of categories and

TABLE 2
Determination of Frequency Values

More low intensity samples			More high intensity samples		
Intensity rank	f	F_i	Intensity rank	f	F_i
A (lowest)	4	1.4	A (lowest)	1	1.0
B	3	2.73	B	2	1.3
C	2	3.9	C	2	2.1
D	2	4.7	D	3	3.26
E (highest)	1	5.0	E (highest)	4	4.6

	Average samples per category									
Category	1	2	3	4	5	1	2	3	4	5
	A: 2.4	A: 1.6	B: 2.2	C: 1.8	D: 1.4	A: 1.0	B: 0.6	C: 0.8	D: 0.8	E: 2.4
		B: 0.8	C: 0.2	D: 0.6	E: 1.0	B: 1.4	C: 1.8	D: 2.2	E: 1.6	
	2.4	2.4	2.4	2.4	2.4	2.4	2.4	2.4	2.4	2.4

Note: Frequency values determined for 12 samples rated on a 5-category rating scale when the attribute of interest is present in all samples but has only 5 different levels. Examples are shown for a greater frequency of low intensity samples (left) and a greater frequency of high intensity samples (right).

samples to be evaluated, the F_i values for each sample can be obtained for each sample independently of the rating session on the basis of a rank ordering. The ordering is based on the physical variable(s) associated with the sensory attribute of interest. A hypothetical example is given in Table 2. In this example, assume that 12 products are being evaluated on a number of attributes using a 5-category rating scale. However, for one of the attributes it is known that the 12 products represent only 5 levels of intensity. The five levels are, from lowest to highest intensity, A (four products), B (three products), C (two products), D (two products), and E (one product). With the total of 12 products and the tendency toward equal category use, it can be expected that each product will be rated on this attribute in each category an average of 2.4 times (12 products/ 5 categories = 2.4). The tendency to match range of products to range of available categories will also ensure proper rank ordering of samples. This knowledge allows calculation of the ratings products would receive on this attribute if frequency were the only influence on ratings. Table 2 shows the procedure schematically. By distributing products equally among the categories and calculating the ratings on the basis of this distribution, the F_i values are obtained. For example, intensity level A, the lowest intensity, is represented four times. On the average, level A would be rated in category 1 about 2.4 times and in category 2 about 1.6 times (2.4 + 1.6 = 4.0) if frequency were the only effective variable. Therefore, F_i for sample A for this attribute is obtained by:

$$2.4 \times 1 = 2.4 \tag{1}$$

$$1.6 \times 2 = 3.2 \tag{2}$$

$$\text{Sum} = 5.6 \tag{3}$$

$$5.6/4 = 1.4 \tag{4}$$

The F_i value for sample A is 1.4. This is the rating that product A would receive on this attribute if *only* relative frequency were the determining factor. Similarly, product B would be rated 2.73, and so on for each product. If the relative frequencies were reversed such that A were the least frequent and E the most frequent, than the F_i values would change accordingly as shown in Table 2.

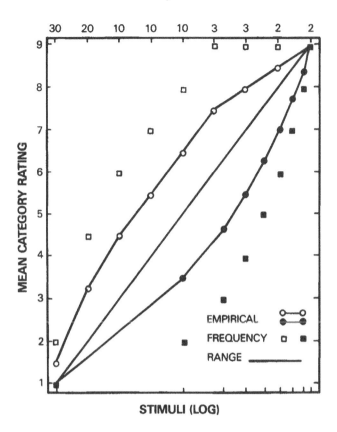

FIGURE 2.   Effects of sample frequency and spacing on category ratings. Samples that occur more often, or are perceptually very close to each other in intensity, tend to receive higher ratings than they otherwise would. (From Parducci, A., *Handbook of Perception*, Vol. 2, Carterette, E. C. and Friedman, M. P., Eds., Academic Press, New York, 1974. With permission.)

In a practical sensory lab, the range of samples can be controlled by maintaining constant high and low anchor samples across studies, but frequency is more difficult to control. Experimental studies have therefore been conducted in which the frequency and range have both been controlled in order to illuminate their joint influence on ratings. Psychologically, a straight line is the relationship that ought to be found between category ratings and sample intensity if the sensations are on an interval scale. Experimentation has found that in fact observers behave as if they are compromising between the tendency to use all categories equally often and the linear relationship between ratings and intensity. The resulting ratings are about halfway between the values of F_i, which are independent of sample range, and ratings that would be independent of frequency effects. Figure 2 shows this relationship.[19] Mathematically, it is simply the average of pure range and pure frequency effect ratings:

$$C_i = (R_i + F_i)/2 \tag{5}$$

where C_i is the actual category rating obtained for a given sample and attribute, R_i is the value the sample would have received if frequency were not affecting the ratings, and F_i is the previously calculated frequency value. Since both C_i and F_i are available, the equation can be solved for R_i, which is a measure of attribute intensity that is independent of the influence of unequal frequencies. Consequently, R_i values can be used to maintain continuity of ratings across studies when frequencies of attribute intensities vary but constant end anchors are maintained. This model accounts for about 80% of the variability in category ratings.[19] It might be noted that an unscrupulous investigator can obtain a great degree of control over sample ratings by simply manipulating the range and frequency of samples to favor his or her pet hypothesis. On the other hand, the conscientious investigator can use this understanding to enhance reliability across studies.

Calculating F_i values depends on knowing the frequencies at which the different attribute intensities occur. In practice, this is not always possible before the study is run. However, it does not matter *when* the F_i values are calculated. They may be determined after data collection is complete, and associated chemical analyses of the products allow good estimation of which samples are perceptually equal on which attributes. An important area for applied research on this issue is the question of how range and frequency of attributes other than the one(s) being rated affect the ratings. This question has not been systematically addressed.

In summary, the values obtained with category rating scales can be strongly influenced by the range and frequency of samples being rated. If this is the case, then the values are on an ordinal scale and inferences about magnitude of differences between samples will be misleading. That is, the rated differences on category scales are relative. A transformation of category ratings to R_i values when possible results in data points that are independent of frequency effects and are probably on an interval scale. These values allow inferences about magnitude of difference between samples. Since the range of samples has a strong effect on ratings, maintaining constant high and low standards across studies will also tend to stabilize ratings. Finally, equally frequent and equally spaced attribute intensities, to the extent they are obtainable, will stabilize the effects of frequency on ratings.

1. Other Influences on Category Ratings

Range and frequency of samples introduce large and unavoidable influences on category ratings.[15,19,23] There are numerous other influences that also deserve mention. These have been discussed in detail elsewhere,[15] and most of the present discussion is drawn from that source.

In the *response equalizing* tendency, the observers distribute ratings over the available range of categories; that is, a small number of samples will tend to be spread out over the entire range of categories, and a large number of samples will be compressed into the range. This will obviously affect the rated differences between the samples. Using an adequate number of categories to cover the range of samples will tend to operate against this influence.

In opposition to the tendency to use all available categories is the tendency to underestimate samples at the high end of the sample range and overestimate those at the low end. Poulton[15] refers to this as *contraction*, while others have used the term *regression* effect.[24] This tendency can be countered by using high and low end anchors and familiarizing the observers with those anchors (or even *all* the samples) before actual ratings are made.

Observers will also tend to rate samples as if they are equally spaced geometrically. This is a special case of the frequency effect. The influence of this tendency can be offset if equal geometric spacing of the samples does in fact exist.[15]

Finally, various idiosyncratic response tendencies can be observed in the use of rating scales. For example, some observers will tend to use only one end of the scale, others will avoid the extremes, etc.[10,25]

2. How Many Categories Is Enough?

Even the most casual perusal of the literature will reveal that the most popular number of categories to use is about seven to nine. Is this the right number? Rather than provide any specific recommendations, a good rule of thumb to guide selection is that the number of categories used should provide a wide enough range to account for the perceptual range covered by the samples. If that range is not covered, true differences between products will be missed or underestimated.

This issue can be conceptualized in terms of how *coarse* or *fine grained* the category rating

NUMBER OF CATEGORIES ON SCALE

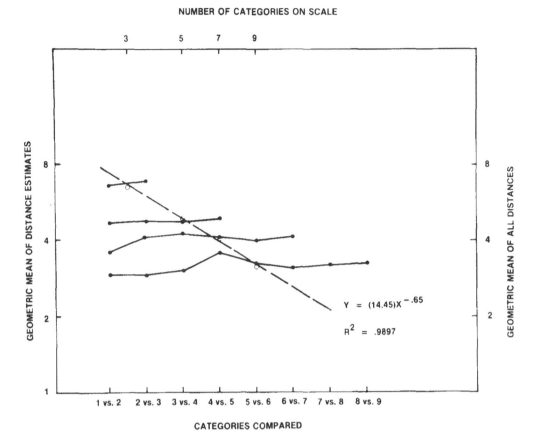

FIGURE 3.   Estimated differences in magnitude between successive categories on scales consisting of three, five, seven, and nine categories. As the number of categories becomes smaller, the estimated difference in magnitude between them becomes larger. Increasing the number of categories results in a "finer grain" rating scale capable of reflecting smaller differences among samples. (From Kuznicki, J. T. and Nagle, D., *Perceptual Motor Skills*, 50, 147, 1980. With permission.)

scale happens to be. The fewer the available categories, the more coarse is the scale. An increase in coarseness means that only a correspondingly large sample difference will be reflected accurately in the ratings. Figure 3 explains why. These data are taken from a study in which subjects were asked to rate the perceptual distances *between* successive categories of category rating scales that varied in number of categories. It can be seen here that, as the number of categories *increases*, the perceptual distance between successive categories *decreases*. Consequently, a given perceived difference between two samples rated on a five-category rating scale might not be large enough to assign the samples to different categories; that is, the difference between samples is less than the distance between successive categories, and the samples are therefore assigned the same rating. However, with more categories available, the distances between successive categories will be smaller than the differences between products. Observers will then assign the products to appropriately different categories. This is not a problem limited to only small differences that have no practical value. This is easily demonstrated using a model system which employs samples having large and obvious differences. Figure 4 and Table 3 show such data. Observers were asked in one part of this study to rate the sweetness of 4, 8, and 16% sucrose solutions on a five- or a nine-point category scale. A separate group of observers was used for each scale. The differences among these samples in sweetness intensity are large, obvious, and nondebatable. On the nine-point category scale, average sweetness ratings between 4 and 8% sucrose and between 8 and 16%

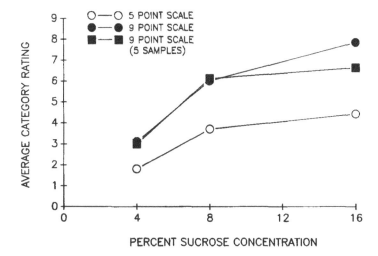

FIGURE 4. Category ratings of the sweetness of the same three concentrations of sucrose solution on five- and nine-category rating scales. Increasing the number of categories increases the values of the ratings and also increases the differences between the ratings of successive samples. Adding more samples (filled squares) decreases the magnitude of the ratings and the differences between successive samples.

TABLE 3
Differences in Category Ratings of Sweetness

Scale used	4.0% Sucrose vs 8.0% sucrose	t	8.0% Sucrose vs 16% sucrose	t
Nine points	2.9	12.806[a]	1.8	5.625[a]
Five points	1.9	8.26[a]	0.7	1.79
Nine points (5 samples)	3.1	3.875[a]	0.5	0.66

[a] Significantly different at confidence level exceeding 95%, N = 30.

sucrose are both statistically significant and the differences are clear in the figure. On the five-category scale, the perceptually very large difference between 8 and 16% sucrose is not reflected in the ratings. In a further experiment in which 2 and 32% sucrose were added as additional samples to be rated on a nine-point scale, the difference between 8 and 16% sucrose again disappears. In this case, the difference disappears because the number of samples was increased, but the size of the scale was not. The conclusion is that if the scale does not allow adequate coverage of the sensory range being studied, or is too coarse, real differences will be missed or underestimated. The chances they will be missed are great, even for very large differences. The effects of number of categories and samples on ratings are called the *category* and *stimulus* effects, respectively, and are described in detail by Parducci.[19]

About how many categories constitute an adequate number most of the time? Teght-soonian[16-18] has reviewed considerable literature showing that, as a general rule, the difference in perceived sensation intensity between the lowest intensity perceptible (i.e., threshold) and the highest intensity perceptible without pain covers a range of about 1.5 log steps for all intensive sensory continua studied. This is referred to as the sensory dynamic range. On this basis, it has been suggested that a category scale with about 33 categories should cover the sensory dynamic range ($\log_{10}33 = 1.5$).[27] Extending the logic further, most practical rating studies do not use samples approaching the high intensity extreme of pain, and threshold data

TABLE 4

Factorial Design Used to Determine the Effects of Five Levels of Sweetener and Three Levels of Citric Acid on the Perceived Orange Flavor Intensity of an Orange Beverage

Citric acid concentration (*M*)	Sucrose concentration (*M*)				
	0.12	0.21	0.30	0.39	0.48
0.003	0.12/0.003	0.21/0.003	0.30/0.003	0.39/0.003	0.48/0.003
0.0125	0.12/0.0125	0.21/0.0125	0.30/0.0125	0.39/0.0125	0.48/0.0125
0.05	0.12/0.05	0.21/0.05	0.30/0.05	0.39/0.05	0.48/0.05

Note: A total of 15 beverages was prepared. Each of the beverages contained one of the five levels of sucrose listed across the top of the table. At each of the five sucrose levels, one product sample contained 0.003 *M* citric acid, one contained 0.0125 *M* citric acid, and one contained 0.05 *M* citric acid. Therefore, each level of sucrose was mixed with each level of citric acid to allow determination of how the two variables interact. Columns in the table represent sucrose concentration, and rows represent citric acid concentration. Each cell represents one of the 15 treatments.

at the low extreme are best collected by other methods. Therefore, the truncated perceptual range usually studied in a sensory lab can almost always be adequately covered by about 20 categories. In fact, this number of categories is in the range where ratings and open-ended methods such as magnitude estimation begin to show exactly congruent results.[28]

3. Functional Measurement

Examples of very successful applications of appropriately and carefully used category rating scales are available and easily found in the literature. A particularly useful application in practical situations is in the evaluation of sensory integration. Most sensations evaluated in chemometrics are multidimensional, that is, the perceived sensory attributes are contributed to by more than one variable and often by many variables. A frequent objective is to determine how these variables are integrated by the sensory systems to result in a more-or-less unitary sensation. This obviously begins to fall into the realm of regression and multidimensional techniques which are indispensable in developing integration models.

Reviews of the types of algebraic models of sensory integration are available, and the use of category rating scales in their development can be found in the same sources.[11,23,29]

A useful way to study sensory integration is with factorial experimental designs in which each level of one variable is represented at each level of a second variable, both of which contribute to the sensation of interest. For example, in an earlier chapter it was demonstrated that the number of observers who perceived an increase in tartness of an orange beverage increased with sourness but decreased with sweetness. By obtaining category ratings of tartness for products appropriately formulated with systematically varying levels of sweetness and sourness, the relative contribution of each of these variables to tartness can be estimated.

Using samples identical to those examined in Chapter 6 to illustrate the use of paired difference tests, a study was run to measure tartness in the orange beverage context. The actual samples and study design are shown in Table 4. Three levels of citric acid (0.003, 0.0125, and 0.05 *M*) were each formulated at 5 levels of sucrose (0.12, 0.21, 0.3, 0.39, and 0.48 *M*) to result in a 3 × 5 factorial design and a total of 15 samples. To adequately cover the perceived sensory range, each of these samples was rated on a 0—20 category rating scale. Before rating, the 30 observers tasted the two extreme samples, i.e., (1) the one containing 0.003 *M* citric acid and 0.12 *M* sucrose at the low end and (2) the one containing 0.05 *M* citric acid and 0.48 *M* sucrose at the high end. They were told these represented the extremes of tartness that would be encountered in this study, and that they would also occur among the set of 15 samples that they would actually rate. Care was taken to adequately clear the mouth and allow recovery from adaptation between sample tastings. The samples were given to each observer in a different random order.

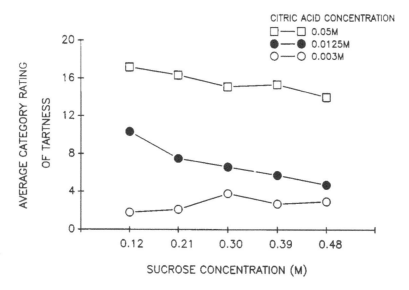

FIGURE 5. Average category ratings of tartness. Each line represents one level of citric acid. Each of the three levels of citric acid was mixed with one level of sucrose (as represented by the X-axis) in a factorial design. Subjects rated the intensity of tartness produced by each mixture. The graph therefore depicts how sweet and sour sensations are integrated into perceived tartness and indicates that sourness of citric acid has a large positive effect while the sweetness of sucrose has a smaller negative influence on ratings.

The results of the tartness ratings are in Figure 5. This figure provides a picture of how sweetness and sourness are integrated into the perception of tartness as defined here. Each line in the figure represents one level of citric acid, and the abscissa represents increasing concentration of sucrose. At the lowest acid level, tartness remains virtually constant as sucrose concentration increases, and therefore sweetness increases. As acid concentration increases, tartness ratings also increase considerably. Tartness at the two highest acid concentrations decreases as sucrose concentration increases, but this effect is clearly not as strong as the increase in tartness with acid level. It can therefore be concluded that sourness has a stronger effect on tartness than does sweetness. In algebraic terms, sourness has a greater and positive weight. Sweetness has a smaller and negative weight, i.e., it subtracts from tartness. A general form of this functional relationship for the two higher acid levels would be:

$$T = w_a A - w_s S \qquad (6)$$

where T is tartness, A is acid level, and S is sucrose level; w_s and w_a are the weights representing the contribution of sucrose and acid to tartness, respectively. This model states that, in general, acid will increase ratings of tartness while sucrose will cause them to decrease. However, Figure 5 shows that, at the lowest acid concentration, tartness does not decrease with increased sweetness, that is, the effect of sucrose depends on which level of acid it is mixed with. This dependence is referred to as an interaction between these two variables, and such a term must be included in the model if accounting for the lowest acid level is desirable. Full development of various kinds of algebraic integration models is discussed in Anderson's text,[11] and fitting specific functions to data is discussed in Chapter 9 of this volume.

B. MAGNITUDE ESTIMATION

Magnitude estimation is a means of scaling perceived sensation intensity by assigning numbers to the sensations in proportion to their intensities. No limit is placed on the range or size of numbers that observers may use. Typical instructions for magnitude estimation are shown in Table 5. In these particular instructions, the observers are instructed specifically to

TABLE 5
Sample Magnitude Estimation Instructions

In this study you will be asked to judge how thick (i.e., viscous) each of a series of beverage samples feels in your mouth. You will do this by assigning numbers to the samples in proportion to their thickness. You will first be given a reference sample. With one sip take all of the reference sample into your mouth, swish it around, then spit it out. Use the number 20 to represent the thickness of this reference sample. For each succeeding sample, assign numbers in proportion to their perceived thickness. If the first sample seems twice as thick as the reference, rate it 40; if it seems half as thick as the reference, rate it 10, and so on. Then, judge each remaining sample in relation to the one rated just before it; that is, rate the second sample relative to the first, the third sample relative to the second, and so forth. You may use any number that seems appropriate to describe the thickness of the sample, including fractions, but not negative numbers. Do not limit yourself to any one range of numbers (e.g., 1 to 9, 0 to 100, etc.). Let the size of the numbers be in proportion to the thickness of the samples, however low or high that might be.

use ratios; that is, if a given sample is rated "10" for a certain attribute and the next sample is perceived as being twice as intense for that attribute, then it should be assigned a "20" and so on. Instructions are sometimes modified to direct observers to simply match the size of the number used to the magnitude of the sensation perceived.[30] The two procedures give very similar results. Numerous other examples of magnitude estimation instructions are available,[10,12,20] and the procedure itself and its history has been extensively researched and described.[10,12,15,20] The method is very popular because it is very easy to use and has the important advantage that, when properly applied, it results in a ratio scale of sensory magnitude. Most physical parameters of sensory stimuli can also be expressed in terms of ratios. This consequently allows the possibility of summarizing data in the form of a power function. Power functions plot as straight lines when the sensation intensity and the physical parameter eliciting the sensation are plotted in log-log coordinates. The exponent of the power function is the slope of the line. The exponent therefore provides a convenient summary index of the data. As it turns out, a wide range of sensory-physical relationships can be described by power functions.[20] However, there is no strict connection between power functions and magnitude estimates. Any function or procedure that appropriately and adequately describes the data may be used with magnitude estimates.

It is sometimes stated that magnitude estimation requires extensive training of observers and is therefore only appropriate for very small studies where one or a few attributes are rated.[9] It is true that some observers show initial discomfort with the relatively free use of numbers, but virtually all become accustomed to it after a few minutes practice. Some observers (about 5%) never become comfortable, but it is not clear whether this has to do with technique itself or product ratings generally.[10] Further, of this 5%, perhaps half of them use the procedure correctly despite their discomfort. The experience of the present author has been very positive, and no need for extensive training has ever been encountered. In addition, data are available showing that magnitude estimation can be used by untrained but properly instructed observers[31,32] while apparently no data is available indicating that extensive training is needed.

As with category rating scales, or with any measurement technique, sensory or physical, the variables which influence the results must be appreciated. Detailed discussions of these are available for magnitude estimation,[10,12,15-18,20] and some of the more important considerations will be reviewed here.

1. The First Sample: The Standard and Modulus

It is sometimes helpful to provide observers with a starting point for magnitude estimation. This is done by presenting them with a standard sample, the intensity of which is preassigned a number by the experimenter. This number is called the *modulus*. For example, in studies of taste and flavor, the observer may be presented with a sample of 10% glucose solution. They are told to taste the sample and attend to its intensity. They are further told that the intensity of this sensation should be assigned a certain number, such as "20". Every sensation experienced from further samples tasted throughout the study is then rated relative to this standard and modulus. This has the advantages of getting observers started, providing a common reference, allowing comparison across different studies which use the same standard, and insuring that numbers used by different observers are in the same range with the consequent reduction in variability of estimates. This latter advantage avoids the need to mathematically adjust the numbers (see below).

In practice, there are different ways to use the standard and modulus.[20] One way is to present it before every sample to ensure it is not forgotten. A variation of this is to make it available at the will of the observer or at certain preselected times throughout the study. An efficient method is to instruct the observers to rate the first sample relative to the standard, and to rate each successive sample relative to the one immediately preceding it. In this way, each sample and its rating becomes the standard and modulus for the next sample. This method preserves successive ratios and reduces the effort of both experimenter and observer. It should be noted that the order of samples after the standard is independently randomized for each observer (as should always be the case). A drawback to using a standard and modulus is that their positions and size will affect the ratings of the observers.[15,34] If the standard is at the high end of the stimulus range (or the modulus is a large number), the ratings will tend to be compressed; if the standard is at the low end of the stimulus range (or the modulus is a small number), the ratings will tend to be expanded. Therefore, a standard in the middle of the range of stimulus intensities being studied and a modulus of around "20" is often recommended.[20] This issue ought to be kept in proper perspective, as the effects of standard and modulus are really quite small when compared to the effects of differences among the samples being rated. In addition, category ratings are sometimes used in the belief that the standard-modulus issue is thereby avoided. This is not the case, as observers will still tend to use varying ranges of the scale, especially as scale size increases.

2. Nonmodulus Magnitude Estimation

The problem of the standard and modulus affecting the range of numbers used by observers can be avoided completely by allowing each observer to assign a number he or she feels comfortable with to the sensation intensities of the first sample. Successive samples are then rated in relation to the immediately preceding sample, and each observer is presented with an independently determined random order of samples. This procedure balances out the effects of the standard, as every sample has a turn at being the sample rated first (as well as occurring in every other position). The only restrictions placed on the first numbers used is that they should have a magnitude which seems to match the intensities of the sensations they are being assigned to. If observers, or the experimenter, feel uncomfortable with this at first, a practice rating on one or two samples will almost always overcome the discomfort. When no modulus is used, variability in the data will come from two major sources. The source of variability that is of interest to the experimenter is the variability due to the real differences in the samples; that is, observers assign numbers to samples to reflect perceived differences among them, and these are the experimental data of interest. A second source of variability, which is not of interest to the experimenter, is due to different observers using different ranges of numbers. For example, one observer may rate sample A as "4" and sample B as "12", while a second observer may rate these same samples as "10" and "30", respectively. Since magnitude estimation is ratio scaling, the important piece of information here is that sample B is perceived to be three times as intense as sample A. This 3:1 ratio can be expressed with numbers of any size. In general, the size of the numbers used by observers is not of interest, but their ratios are.

The variability due to arbitrary choice of number size can be eliminated from the data without changing the ratios among sample ratings. This is very easily done by multiplying each observer's data points by a constant chosen to equate the average ratings of each observer. The original ratios can be preserved and the average rating of each observer equated to any number desired by the formula:

$$ME' = ME_{si}(C/X_{si}) \tag{7}$$

This operation is referred to as calibrating the magnitude estimates. ME' is the calibrated estimate, C is the value to which each observer's *average* rating is to be equated, ME_{si} is a magnitude estimate from the observer whose estimates are being calibrated, and X_{si} is the average of the uncalibrated estimates from that same subject. It can be seen from the formula that if a given subject has, on the average, used numbers larger than the value of C, that subject's data points will all be multiplied by a number less than one and the observer's calibrated estimates will therefore all be decreased by an amount which will make their new average equal to C. Similarly, if an observer tended to use very small numbers which averaged less than C, then that observer's estimates will all be multiplied by a value greater than one which will increase the calibrated estimates so that their average equals C. A worked example is shown in Table 6 and demonstrates that the procedure equates averages without altering ratios. In this example, the value selected for C was the overall average of all data points. When nonmodulus magnitude estimation is used, the raw estimates are calibrated before any other analyses are carried out.

The choice of the value of C is truly arbitrary, but reasonable guidelines for data presentation ought to be followed. For example, if the value of C is chosen to be 20, then the resulting calibrated data will be easier to plot, and the graphs will be visually more appealing than if a very large or very small value were chosen. One common procedure is to make C equal to the average of the raw magnitude estimates taken across all observers, samples, and attributes. This makes each observer's average equal to the overall average of the raw estimates. Additional procedures which allow anchoring of the calibrated estimates to verbal descriptors of intensities, such as "very intense", "moderately intense", etc., are also used[10] and have the advantage of indicating the absolute, as well as relative, intensities of sample attributes. In this case, subjects are asked at the end of a study to tell what number they would have used if a product had a "very intense" level of an attribute, "moderately intense" level, etc. These numbers are then used as a basis for calibrating the raw data. The only restriction is that the rank order of the estimates applied to the verbal anchors be the same as the rank orders of the intensities implied by the verbal anchors. For example, the rating of the verbal anchor for "moderate intensity" cannot be higher than the rating for the anchor "very intense", etc. A worked example of this procedure is also shown in Table 6.

3. Measures of Central Tendency

Magnitude estimates can be summarized by the geometric mean because they have ratio properties. One advantage of the geometric mean is that it is relatively uninfluenced by extreme values. This is important because magnitude estimates for any given attribute intensity tend to be log-normally distributed, i.e., their distribution tends to have a long tail at the high end.[34] Unfortunately, the geometric mean cannot be determined if any values of zero are obtained. Thus, if a given sample is rated as zero on even just one attribute, the geometric means cannot be determined for that attribute. If the number of zeros is small, various techniques for estimating missing data might be employed to obtain a nonzero value, but the data are not missing in the true sense of the word. Fortunately, the arithmetic mean as well as the median are both also appropriate measures of central tendency for magnitude estimates, and the investigator can safely and confidently use either one.[20]

4. Range and Frequency of Samples

Magnitude estimation is subject to the same range and frequency effects as are category

TABLE 6
Methods of Calibrating Magnitude Estimates

Method 1: Equating each subject's average to the overall average
  A. Compute the average of all data points (i.e., this is the average of every data point from every subject)
  B. Compute the average of *each* subject (i.e., there should be one average for each subject in the study)
  C. Determine this ratio for *each* subject: overall $\overline{X}/\overline{X}_{subject\,i}$; this ratio is determined separately for each subject and is the individual calibration factor for the subject
  D. Each individual subject's data points are multiplied by the calibration factor determined for that subject in step C; the resulting numbers are the calibrated data and each subjects average rating will now equal the overall average, but the ratios among the calibrated ratings will be equal to the ratios among the raw data points

Worked example:

Sample	Raw data	
	Subject 1	Subject 2
A	12	115
B	27	198
C	43	300
Overall $\overline{X}$ =		115.83
$\overline{X}_{subject\,1}$ =		27.33
$\overline{X}_{subject\,2}$ =		204.33

Subject 1 calibrating factor = 115.83/27.33 = 4.23
Subject 2 calibrating factor = 115.83/204.33 = 0.56

Calibrated data*

Sample	Calibrated data	
	Subject 1	Subject 2
A	50.76	64.4
B	114.21	110.8
C	181.89	168.0

*Subject 1's data points have been multiplied by 4.23; Subject 2's data points have been multiplied by 0.56; ratios among calibrated data points remain the same as among raw data points.

Method 2: Calibrating against verbal intensity anchors
  A. Compute the average of each subject's estimates for the verbal anchors:

Verbal intensity anchor	Estimate
Very great deal	140
Very much	100
Moderate	75
Slight	40
Average	88.75

  This is done for every subject individually.
  B. Compute the percentage that each of the intensity phrases (i.e., very great deal, etc.) is of the average obtained in step A.
  That is, 88.75 is the calibration factor for this subject:
$$(140/88.75) \times 100 \quad \text{very great deal is} \quad 157.74\% \text{ of } 88.75$$
$$(100/88.75) \times 100 \quad \text{very much} \quad \text{is} \quad 112.67\% \text{ of } 88.75$$
$$(75/88.75) \times 100 \quad \text{moderate} \quad \text{is} \quad 84.50\% \text{ of } 88.75$$
$$(49/88.75) \times 100 \quad \text{slight} \quad \text{is} \quad 55.21\% \text{ of } 88.75$$
  C. Calibrate this subject's raw data using this same calibration factor of 88.75. For example, if this subject's raw magnitude estimate of the moistness of a given product was 48, then this subject's calibrated estimate is $(48/88.75) \times 100 = 54.08$. According to the verbal anchors, this subject meant to indicate that the product's moistness was what the subject considered to be slightly moist (slight = 55.21).
  D. Repeat this procedure for each individual subject before computing summary statistics on the entire data set.

ratings. The effects are much smaller with magnitude estimation than with category ratings,[16-18,20,35] and are not always observed.[36] The general effect of stimulus range is that the slope of the psychophysical function will decrease somewhat as the range of stimuli for a given attribute increases.[18,37] Increased sample frequency results in a *local* steepening of the function in the area of higher frequency, but does not affect the overall result,[36] and the effect is much less potent than the frequency effect seen with category ratings. Since magnitude estimation does not limit the size of the scale that observers use, the issue of scale size (i.e., the category effect) does not exist as it does with category scale ratings.[21]

C. VARIATIONS OF MAGNITUDE ESTIMATION

Several varieties of magnitude estimation exist. These are detailed extensively in several sources,[10,20,30,38] and will only be briefly described here. In *magnitude production*, the observer is presented with a number and asked to adjust the intensity of a sensation to match the size of the number. This is exactly the opposite of magnitude estimation and is easily used with visual and auditory stimuli that can be conveniently and continuously adjusted by observers. Chemical stimuli are too difficult to adjust during a rating session to allow extensive practical use of the method for chemometric applications. *Cross-modality matching* involves the adjusting of the perceived intensity of one sensation to match the intensity of another sensation from a different modality. Observers find this task to be very natural and even easier than direct magnitude estimation. An important application in chemometrics is in time-intensity profiling, where an easily controlled and continuously adjustable variable like the intensity of an auditory stimulus can be adjusted to match the rise and decline in taste intensity over time.

Magnitude matching is a relatively new procedure[39,40] which is derived from and somewhat similar to cross-modality matching. In magnitude matching, subjects are presented with stimuli alternating between two different modalities and attempt to rate all the stimuli on one scale of sensory magnitude using either magnitude estimation or category ratings.[40] By using one of the judged modalities as a standard for comparison and assuming that all subjects perceive it equally, comparison of perceived sensory magnitude can be made *across* individuals as well as within individuals.

Consider, for example, the typical rating session in which each observer rates several samples on various attributes. When only one continuum is rated (such as taste intensity), the results will show systematic differences in ratings of the samples and most observers will show very similar results in a well-run study; that is, we can determine if a given observer finds one sample more sweet than another, and if his or her judgment agrees with the rating of another observer. However, when two observers taste the same sample and give it the same rating, are they really experiencing the same taste intensity, or is their use of the same rating fortuitous? Most people who work with flavors have ample anecdotal evidence to suggest that not everyone experiences flavor alike, and we know conclusively that both genetic factors[41] and aging[42-44] affect taste and smell responsiveness as measured by threshold methods. Magnitude matching allows direct comparison of the perception of suprathreshold sensations across different observers.

Magnitude matching assumes that all observers in a study are about equally responsive to one of the modalities being used. Ratings from the second modality are then calibrated using a calibration factor determined from the ratings of the first modality. For example, suppose two groups of people are tested using magnitude matching of the bitter taste of 6-*n*-propylthiouracil (PROP) and sound intensity of a 1000-Hz tone. It is known that sensitivity to PROP (and other bitters) is genetically determined with so-called tasters having a much lower threshold than nontasters.[41] One group of observers in the study is classified as PROP tasters, and the other group consists of nontasters. After the rating of both sound and PROP stimuli in alternating order (i.e., sound-PROP-sound-PROP, etc.) is complete for both groups, the data are calibrated as follows. Judgments of sound are separated from the data, and a multiplicative factor needed to equate each observer's average sound intensity rating to a common value is determined using the procedure for calibrating magnitude estimates described earlier. This multiplicative factor, based only on sound judgments, is then used to multiply all of that subject's data for *both* sound and taste. This is done for each subject individually before any further data analyses.

Several experiments of the type just described have been conducted.[43-46] These studies have found, for example, that a given suprathreshold concentration of PROP tastes stronger to genetic tasters than nontasters, that PROP tasters also perceive equal concentrations of sucrose and neohesperidin dihydrochalcone to be sweeter than nontasters perceive them, and that different taste qualities are affected differently by aging.

The importance of this technique is very clear. It allows the direct comparison of sensory magnitude *between* individuals via a very simple procedure. The technique, therefore, ought to prove valuable in screening taste panel members for equal sensitivity and also for identifying unusual variations in sensitivity to interesting stimuli. So far, the matching continua that have been used successfully to demonstrate individual differences in taste and smell sensitivity include sound intensity, taste intensity of NaCl, and heaviness of lifted weights.[39-46] The main criterion for the matching continuum is that subjects be known to be more uniform in their responsiveness to it than to the test continuum.[40] If this assumption is met or approximated, then magnitude matching will provide reliable results.

REFERENCES

1. **Stevens, S. S.,** On the theory of scales of measurement, *Science,* 103, 677, 1946.
2. **Stevens, S. S.,** Mathematics, measurement, and psychophysics, in *Handbook of Experimental Psychology,* Stevens, S. S. Ed., John Wiley & Sons, New York, 1951.
3. **Torgerson, W.,** *Theory and Method of Scaling,* John Wiley & Sons, New York, 1958.
4. **Campbell, N. R.,** *An Account of the Principles of Measurement and Calculation,* Longmans Green, London, 1928.
5. **Stine, W. W.,** Meaningful inference: the role of measurement in statistics, *Psychol. Bull.,* 105, 147, 1989.
6. **Davison, M. L. and Sharma, A. R.,** Parametric statistics and levels of measurement, *Psychol. Bull.,* 104, 137, 1988.
7. **Maxwell, S. E. and Delaney, H. D.,** Measurement and statistics: an examination of construct validity, *Psychol. Bull.,* 97, 85, 1985.
8. **Townsend, J. T. and Ashby, F. G.,** Measurement scales and statistics: the misconception misconceived, *Psychol. Bull.,* 96, 394, 1984.
9. **Meilgaard, M., Civille, G. V., and Carr, B. T.,** *Sensory Evaluation Techniques,* Vol. 1, CRC Press, Boca Raton, FL, 1987.
10. **Moskowitz, H. R.,** *Product Testing and Sensory Evaluation of Foods,* Food & Nutrition Press, Westport, CT, 1983.
11. **Anderson, N. H.,** *Foundations of Information Integration Theory,* Academic Press, New York, 1981.
12. **Stevens, S. S.,** *Psychophysics: An Introduction to its Perceptual, Neural, and Social Prospects,* John Wiley & Sons, New York, 1975.
13. **Birnbaum, M. H. and Veit, C. T.,** Scale convergence as a criterion for rescaling: information integration with difference, ratio, and averaging tasks. *Percept. Psychophys.,* 15, 7, 1974.
14. **Birnbaum, M. H.,** Comparison of two theories of "ratio" and "difference" judgments, *J. Exp. Psychol.: Gen.,* 109, 304, 1980.
15. **Poulton, E. C.,** Models for biases in judging sensory magnitude, *Psychol. Bull.,* 86, 777, 1979.
16. **Teghtsoonian, R.,** Range effects in psychophysical scaling and a revision of Steven's law, *Am. J. Psychol.,* 86, 3, 1973.
17. **Teghtsoonian, R.,** On the exponents in Steven's law and the constant in Ekman's law, *Psychol. Rev.,* 78, 71, 1971.
18. **Teghtsoonian, R. and Teghtsoonian, M.,** Range and regression effects in magnitude scaling, *Percept. Psychophys.,* 24, 305, 1978.

19. **Parducci, A.,** Contextual effects: a range-frequency analysis, in *Handbook of Perception*, Vol. 3, Carterette, E. C. and Friedman, M. P., Eds., Academic Press, New York, 1974.

20. **Marks, L. E.,** *Sensory Processes: The New Psychophysics*, Academic Press, New York, 1974.

21. **Parducci, A. and Weddel, D. H.,** The category effect with rating scales: number of categories, number of stimuli, and the method of presentation, *J. Exp. Psychol.: Hum. Percept. Perform.*, 12, 496, 1986.

22. **Riskey, D. R.,** Use and abuse of category scales in sensory measurement, *J. Sens. Studies*, 1, 217, 1986,

23. **Anderson, N. H.,** On the role of context effects in psychophysical judgment, *Psychol. Rev.*, 82, 462, 1975.

24. **Stevens, S. S. and Greenbaum, H.,** Regression effects in psychophysical judgment, *Percept. Psychophys.*, 1, 439, 1966.

25. **Moskowitz, H.,** Psychophysical and psychometric approaches to sensory evaluation, *Crit. Rev. Food Sci. Nutr.*, 9, 41, 1977.

26. **Kuznicki, J. T. and Nagle, D.,** Magnitude estimates of distances along category scales, *Percept. Mot. Skills*, 50, 147, 1980.

27. **Foley, H. J., Cross, D. V., Foley, M. A., and Reeder, R.,** Stimulus range, number of categories, and the "virtual" exponent, *Percept. Psychophys.*, 34, 505, 1983.

28. **Gibson, R. H. and Tomko, D. L.,** The relation between category and magnitude estimates of tactile intensity, *Percept. Psychophys.*, 12, 135, 1972.

29. **Anderson, N. H.,** Functional measurement and psychophysical judgment, *Psychol. Rev.*, 77, 153, 1970.

30. **Zwislocki, J. J. and Goodman, D. A.,** Absolute scaling of sensory magnitude: a validation, *Percept. Psychophys.*, 28, 28, 1980.

31. **Stevens, J. C. and Tulving, E.,** Estimations of loudness by a group of untrained observers, *Am. J. Psychol.*, 70, 600, 1957.

32. **Lawless, H. T. and Malone, G. J.,** A comparison of rating scales: sensitivity, replicates and relative measurement, *J. Sen. Studies*, 1, 155, 1986.

33. **Engen, T. and Levy, N.,** The influence of standards on psychophysical judgments, *Percept. Mot. Skills*, 5, 193, 1955.

34. **Stevens, J. C.,** A Comparison of Ratio Scales for the Loudness of White Noise and the Brightness of White light, dissertation, Harvard University, Cambridge, MA, 1957.

35. **Marks, L. E., Szczesiul, R., and Ohlott, P.,** On the cross-modal perception of intensity, *J. Exp. Psychol.: Hum. Percept. Perform.*, 12, 517, 1986,

36. **Pradhan, P. L. and Hoffman, P. J.,** Effect of spacing and range of stimuli on magnitude estimation, *J. Exp. Psychol.*, 66, 533, 1963.

37. **Stevens, J. C.,** Stimulus spacing and the judgment of loudness, *J. Exp. Psychol.*, 56, 246, 1958.

38. **Poulton, E. C.,** The new psychophysics: six models of magnitude estimation, *Psychol. Bull.*, 69, 1, 1968.

39. **Stevens, J. C. and Marks, L. E.,** Cross-modality matching functions generated by magnitude estimation, *Percept. Psychophys.*, 27, 379, 1980.

40. **Marks, L. E., Stevens, J. C., Bartoshuk, L. M., Gent, J. F., Rifkin, B., and Stone, V. K.,** Magnitude-matching: the measurement of taste and smell, *Chem. Senses*, 13, 63, 1988.

41. **Bartoshuk, L. M.,** Bitter taste of saccharin related to the genetic ability to taste the bitter substance 6-*n*-propylthiouracil, *Science*, 205, 934, 1979.

42. **Murphy, C.,** Taste and smell in the elderly, in *Clinical Measurement of Taste and Smell*, Meiselman, H. L. and Rivlin, R. S., Eds., Macmillan, New York, 1986.

43. **Stevens, J. C., Bartoshuk, L. M., and Cain, W. S.,** Chemical senses and aging: taste *versus* smell, *Chem. Senses*, 9, 167, 1984.

44. **Bartoshuk, L. M., Rifkin, B., Marks, L. E., and Bars, P.,** Taste and aging, *J. Gerontol.*, 41, 51, 1986.

45. **Murphy, C. and Gilmore, M.,** Quality-specific effects of aging on the human taste system, *Percept. Psychophys.*, 45, 121, 1989.

46. **Gent, J. F. and Bartoshuk, L. M.,** Sweetness of sucrose, neo-hesperidin dihydrochalcone, and saccharin is related to genetic ability to taste the bitter substance 6-*n*-propylthiouracil, *Chem. Senses*, 7, 265, 1983.

Chapter 8

COLLECTION OF SENSORY AND CHEMICAL DATA FOR TEA

I. COLLECTION OF SENSORY DATA

This section describes how a set of sensory data was collected for analysis. The objective of this collection was to obtain a common data set to be discussed in later chapters describing two different types of multivariate techniques: discriminant analysis and factor analysis. Both of these procedures can be carried out on the same data and provide slightly different descriptions of it. In discriminant analysis, samples that are rated in a sensory evaluation session are classified in terms of whether subsets of samples have enough in common to be included in the same group. Conversely, the descriptor variables used to describe the samples could also be analyzed by that technique to determine if any set of descriptors is so close in meaning to the subjects doing the ratings that they can profitably be considered as identical. This is an all-or-none decision. In factor analysis, each variable is described in terms of how much in common it has with each of several groupings typically identified. This is a continuous procedure indicating degree of relationship. While these procedures certainly do not exhaust the domain of multivariate techniques, they are representative of it and both are used commonly enough that they provide a very good practical basis to develop a general feeling of what these sorts of techniques can do. Our purpose is not to provide new learning about tea, but rather to use the data as a means of demonstrating statistical methods.

A. SELECTION OF SAMPLES

Orange-flavored teas were selected for evaluation. These products were selected because an investigation of the types of orange teas on the market indicated that a relatively small number of teas, five, could provide a very representative sampling of widely available brands of orange tea. Sampling the teas casually before formal evaluation indicated that the five varieties provided an excellent range of orange, tea, and spice flavors. Before the formal rating session, several pilot tastings were conducted among small groups to ensure an adequate range of flavor intensities for allowing existing correlations to appear. If all teas are at the same total intensity levels overall, and many attributes are at about the same taste intensity levels, the range of values obtained in the ratings will be restricted. Restricting the range of values will result in a failure to find significant correlations that actually do exist (see Chapter 9).

The results of the pilot tastings were also used to develop brewing procedures for the teas. First, it was found that if all five teas were prepared by brewing two tea bags in 500 cc of boiling water for 3 min, the naturally occurring range of flavor intensities was quite large. Consequently, that was selected as a standard brewing condition. Next, brewing procedures were selected for high and low references which would help the raters anchor the high and low ends of their rating scales. One of the teas, sample L, was found to be quite intense at the standard brewing strength and was therefore selected as the "high" reference sample. Sample L was also prepared by brewing one tea bag for 1 min in 500 cc of boiling water to provide the "low" reference sample. The low reference was labeled "M". Finally, sample L was prepared at a strength of two tea bags brewed for 1 min in 500 cc of boiling water to provide a sample judged to be near the middle of the intensity range of all other samples rated. This middle sample was labeled "T". The brewing procedures therefore provided for a set of seven samples preselected to adequately represent the quality domain of orange teas and also to provide a wide range of sample intensities. Before being tasted by the raters, the samples were allowed to cool to about 130°F and approximately 15 cc of each sample was provided to each of 26 raters. The samples were given to each of the raters in an independently determined random order.

TABLE 1
Instructions Given to Subjects Before Tea Sensory Evaluation Session

Orange and Spice Tea Study Instructions

We would like you to rate the intensities of several attributes that are associated with orange and spice tea. The specific interest of this study is to determine how these attributes change systematically over a wide range of formulations. Your rating will provide us with important information on how people perceive these samples. In order for us to obtain all the information needed, we would like you to rate seven samples of an orange and spice tea.

While rating the samples, follow these instructions:

1. Please clear your mouth *before* tasting each sample by eating two oyster crackers and rinsing thoroughly with the water provided in the pitcher. You should repeat this procedure *before* tasting each sample.
2. Smell the sample and rate the aroma.
3. Now place all of the sample in your mouth but do not swallow it. The teas are all edible with nothing in them to harm you. However, swallowing them will make it very difficult for you to rinse your mouth thoroughly because some of the tea will remain in your throat and the back of your mouth.
4. Hold the sample in your mouth for a few seconds and then spit it into the sink. Then rate it in the appropriate spaces.
5. While going through the study, please concentrate on each sample separately and rate it carefully.

Repeat this procedure for all the samples.

B. THE RATERS

The individuals who rate the samples are variously referred to as respondents, subjects, raters, panelists, etc., with the preferred term seemingly related to the background of the experimenter. Those with a food science background often use the term "panelist", while those with a sensory psychology background tend to use the term "subjects" which is customary in that field. In any event, the terms here all mean the same thing, i.e., the people who rate the samples, and they are (and throughout this book) used interchangeably here.

There are various types of subjects that can be selected to rate products. For example, very often ratings are obtained from small groups of highly trained individuals who spend a good portion of their working time doing such ratings.[1] These ratings tend to be highly reliable even with only a few raters, and the number of flavor nuances that can be discerned and described by such groups is often much larger than can be obtained from naive consumers. Such groups can be valuable, but are often not available, can be time consuming to train, and very difficult to maintain. Finally, the high degree of sensory discrimination powers which these groups can develop over time makes it difficult to build a needed bridge between them and the typical consumer who usually can be expected to reliably discern fewer sensory attributes. For these reasons, the subjects in the sensory study described here were chosen from the untrained employee pool of sensory subjects. These subjects tended, for the most part, to be experienced at sensory ratings, but were never given any formal descriptive panel training. The data obtained from such subjects can be expected to be much closer to the average naive consumer data than that obtained from a highly trained group. The present authors have had a high degree of success using such groups, and their usefulness and value has been noted by others working on similar problems.[2-4]

The instructions given to the subjects immediately before they received the samples for rating and the actual rating sheet and rating instructions are shown in Tables 1 and 2, respectively. These instructions describe the details of the sensory evaluation session. In addition, the ratings were conducted in individual sensory rating booths. The rinse water was distilled. The procedures were chosen on the basis of the discussion in Chapter 7 on sensory scaling techniques. It was decided to use a 20-point category scale to allow adequate room for the ratings to reflect differences among the samples. It was therefore necessary to have the subjects first taste a high and low standard as a means of reducing the tendencies of subjects to contract the range of their ratings as discussed earlier. This is done to help ensure an

TABLE 2
Rating Sheet Given to Subjects for Tea Sensory Evaluation

You have been given two reference samples and seven test samples to rate. Smell and taste each of the reference samples first. They represent the LOWEST and HIGHEST *overall* aroma and taste intensities in the test samples. While evaluating the reference samples, establish in your mind a rating for each reference on a 0 to 20 scale where *0 = none* and *20 = very high intensity.* Your ratings for the seven test samples should fall within the range you've assigned for the references. The reference samples are repeated among your seven test samples.

Please evaluate aroma and taste for each tea BEFORE going on to the next one. Take your time and do this very carefully.

Remember to clear your mouth with crackers and rinse with water *BEFORE* tasting each sample.

Assign an intensity rating for each attribute listed below using a rating scale of 0 to 20. (0 = NONE and 20 = VERY MUCH)

TASTE SAMPLES IN THIS ORDER ___   ___   ___   ___   ___   ___   ___

Aroma in the Cup
    Orange
    Clove
    Cinnamon
    Other

Taste Attributes
    Sweet
    Sour
    Bitter
    Astringent (mouth-drying)
    Tea-like flavor
    Candy-like orange flavor
    Fresh orange flavor
    Other fruit flavors
    Clove
    Cinnamon
    Other spice flavor
    Grassy (hay-like)
    Woody
    Other
Overall preference:
    0 = *dislike* very much,
    20 = *like* very much)

adequate spread of the data points to allow existing correlations to appear. Finally, the analyses to be performed on the data, discriminant and factor analysis, use correlations virtually exclusively. Correlations, as discussed in conjunction with regression analysis, are determined by relative values. Since category rating scales can be expected to reflect accurately the relative intensities of the attributes in the products, no attempt was made to determine a "true" psychophysical value, such as what range values[5] may provide. Rather, it was assumed that 20 categories were sufficient to provide data that probably had interval properties and the relative differences in the ratings mirrored accurately the relative differences among the products. Analyses were therefore conducted on the data at face value and their relative nature was accepted.

C. ATTRIBUTES RATED

The attributes rated are shown in Table 2. These attributes were selected with the knowledge that the subjects doing the ratings were not highly trained in descriptive methods. Rather, they were close to naive consumers in terms of their experience in sensory ratings. Consequently, no attempt was made to obtain ratings of subtle flavor nuances that might be discriminated by highly trained individuals but have little meaning to the typical consumer.

The actual attributes rated were determined on the basis of tasting the samples, knowledge of their composition, and experience with previous studies of tea flavor.

II. COLLECTION OF CHEMICAL DATA

The tea samples used in this database have three characteristic flavor properties. They are (1) black tea, typically characterized by a green, grassy aroma and an astringent, sometimes bitter taste, (2) orange flavor probably derived from orange oils and/or essences and the sour or acidic character of citrus products, and (3) spice flavor described as either clove and/or cinnamon.

The aroma character of these three flavors is fairly well characterized. Many of the compounds that contribute to the impressions are known. They can best be analyzed by capillary column gas chromatography. Headspace analyses would yield information from the most volatile to the medium volatility compounds. Extraction procedures would yield information on all but the most volatile compounds. Both types of data were acquired to determine which was most appropriate for the task at hand. Initially, it was felt that both types would have to be combined to obtain a complete volatile profile. The expected range of compound polarities could be handled by a column of low to intermediate polarity.

The taste portion of the flavor is probably derived from the polyphenolic compounds present in the tea and simple organic acids, either indigenous or added to provide a slight sourness to accompany the citrus flavors. Low levels of sugars are present in black teas. The possibility also exists for the presence of added sugars to sweeten the products. All these classes of compounds can be analyzed by high performance liquid chromatography (HPLC). However, the significant differences in chemical characteristics require different sample preparation, analysis conditions, and instrumental detectors. Each class of compounds was analyzed separately.

A. SAMPLE ANALYSES

The tea samples were prepared according to the directions given above. Samples were prepared and analyzed on four separate occasions to minimize the number of samples for each batch of analyses and to mimic the conditions and variations encountered in the preparation of multiple samples for sensory testing. This approach includes the variations in the teas defined by the manufacturing and formulation, as well as the variability resulting from batch preparation. The brew solids (extractable tea solids) for each batch were determined by standard refractive index measurements to identify any gross errors in sample preparation; none were observed.

Triplicate analyses for the second batch of samples were performed to demonstrate the reproducibility of the analytical procedures. For this part, separate samples were taken from a common pot of tea and analyzed independently. The replicates were included in the entire data set.

The analytical procedures were previously developed and validated. Recoveries and detector response factors for individual compounds were determined for the SPE extraction procedure and the flame ionization detector (FID), respectively. The GC procedures used internal standard calibration while the HPLC procedures used external standard calibration. Many peaks were obtained in the GC profiles and the HPLC polyphenolic profile. Since the purpose of this database was to demonstrate concepts, only known compounds for which calibrations had been obtained during method validation were used for the data analyses. It is realized that such data editing risks the loss of potentially important information. However, it does make the presentation of concepts and results more manageable. A rigorous analysis would include all the unknown compounds in the profiles as well. In addition, no data for sugars were included as the levels were very low and did not vary significantly.

The instruments used for gathering the data were interfaced to a multiuser, multitasking laboratory automation computer system (LAS). Digitized signals from the instruments were integrated by the LAS system and the results stored in separate files for each chromatogram. The database was assembled piecewise by matching the peaks from each analysis procedure for all the samples. This yielded a separate database file for the GC headspace, GC extraction, HPLC polyphenolic, and the HPLC organic acids profiles. These databases were then combined to yield the complete chromatographic profile of the samples.

1. Headspace Analysis

Static headspace analyses were performed on the tea samples to determine the presence of extremely light compounds. 2-Pentanone was used as the internal standard. The sample (5 ml) was placed in a 10-ml headspace vial. The sample was spiked with 25 μg of the internal standard yielding a concentration of 5 ppm. The vial was sealed and thermostated at 40°C. An automatic sampler was used to inject the headspace sample into the GC. The sample was cryogenically focused on the head of the column and analyzed under the following conditions:

1. Injection port temperature: 200°C
2. Injection: 1 ml, splitless; syringe temperature 60°C
3. Capillary column: DB-1, 30 m × 0.53 mm id, 1.5-μ phase thickness
4. Column oven temperature profile: Initial temperature = –49°C, hold for 1 min; ramp at 25°C/min to 0°C; ramp at 5°C/min to 150°C; ramp at 25°C/min to 200°C, hold for 5 min.
5. Carrier gas: helium at a flow rate of 6.7 ml/min at 65°C.
6. Detector: FID operated at 275°C

Figure 1a shows a typical headspace profile obtained for one of the tea samples. No extremely light compounds were found in the profile. The first compound of reproducible magnitude in the headspace profile was ethyl butyrate. Peak areas were ratioed to that of the internal standard.

2. SPE Extraction/GC Analyses

Reverse phase SPE columns were used to extract the tea samples for GC analysis. Two columns (J. T. Baker #7070) were used in series. The columns were conditioned for analysis according to the manufacturer's instructions using 8 ml of acetonitrile followed by 10 ml of deionized water. Cyclohexylcyclohexanone was used as the internal standard (spike level 1 ppm), since it did not coelute with any analytes and was well behaved on the SPE columns. The spiked tea (50 ml) was passed through the SPE columns to trap the volatile compounds. The eluate was virtually odorless, indicating that most of the aroma compounds had been removed from the sample. The extract was eluted from the columns with three 2-ml aliquots of methylene chloride. The extract was concentrated prior to analysis by the indirect blow down method described in Chapter 2. The volume (500 μl) was reduced to 100 μl in a conical, limited-volume sample vial. The overall concentration enhancement was approximately 40 to 1.

The concentrated extract was injected into the GC and analyzed under the following conditions:

1. Injection Port Temperature: 225°C
2. Injection: 1 μl, splitless
3. Capillary column: DB-5, 60 meter × 0.25mm id., 1-μ film thickness
4. Column oven temperature profile: 50°C, hold for 5 min; ramp at 5°C/min to 250°C, hold for 15 min.
5. Carrier gas: helium at a flow rate of 1.5 ml/min at 100°C.
6. Detector: FID operated at 275°C

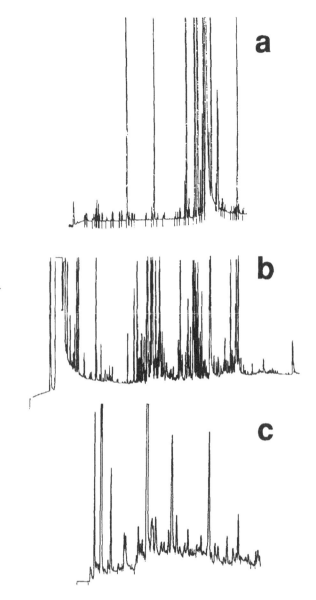

FIGURE 1.   Typical chromatographic profiles for the tea samples. (a) GC-headspace profile, (b) GC-SPE extraction profile, and (c) HPLC polyphenolic profile.

A typical chromatogram of an SPE extract is shown in figure 1b. A total of 123 peaks was obtained in the union of the data from all the samples; there were 22 known compounds. Identity was based on retention times by comparison with standards of the compounds. No further attempt was made to identify the other unknown compounds for this study.

The parts per million (ppm) levels of these 22 compounds were calculated by ratioing the peak areas to that of the internal standard and using previously determined extraction efficiencies and FID relative response factors. The range of levels found in these samples is given in Table 3.

Principal component analysis (PCA) of the headspace and SPE profiles showed a high degree of correlation between the results for the two procedures. It was determined that the SPE extraction data would be used for the volatile characterization, as it could adequately monitor the lightest compound (ethyl butyrate) and provide data on a wider range of compounds.

TABLE 3
Ranges of Levels for Compounds Used to Characterize Tea Samples

SPE/GC		HPLC	
Compound	Range (ppm)	Compound	Range (ppm)
Ethyl butryate	0.00—0.35	Gallic acid	8.68—27.9
α-Pinene	0.00—0.06	Theobromine	4.67—15.3
β-Myrcene	0.00—0.18	Theophylline	0.35—2.06
Ethyl hexanoate	0.00—0.07	EGC[a]	4.79—15.5
Octanal	0.00—0.05	Caffeine	56.9—161
Limonene	0.05—19.7	Epicatechin	1.89—24.7
γ-Terpinene	0.12—0.35	EGCG[b]	14.3—156
Linalool	0.23—1.67	ECG[c]	25.7—147
Cineole	0.10—0.13		
4-Terpineol	0.04—0.10	Citric acid	16.3—72.3
α-Terpineol	0.03—0.80	Malic acid	9.03—19.0
Neral	0.00—0.18	Succinic acid	202—709
d-Carvone	0.02—0.10	Fumaric acid	0.21—0.54
Geranial	0.02—0.23		
Cinnamic aldehyde	0.08—0.26		
Perialdehyde	0.04—0.13		
Undecyclic aldehyde	0.04—0.09		
Neryl acetate	0.00—0.05		
Eugenol	2.08—23.0		
Dodecanal	0.00—0.36		
β-Caryophyllene	0.02—2.62		
Valencene	0.00—1.73		

[a] EGC = epigallocatechin.
[b] EGCG = epigallocatechin gallate.
[c] ECG = epicatechin gallate.

3. Polyphenol Profile

The polyphenols constitute the major group of compounds in tea. Within this class, the catechin flavanols are the most common. Even though many phenolic compounds have been isolated and identified in black tea, standards for most of them are not available. Accurate quantitation is limited to those few where standards are avaialble. It is possible to estimate the levels of compounds for which standards are not available by assuming similar equimolar response factors for compounds containing the same chromophoric groups.

The polyphenolic profiles for the tea samples were determined by gradient elution, reverse phase HPLC with UV detection. Samples were prepared for analysis by filtration through a 0.45-μ membrane filter (Millipore # HA) and diluted 1:1 with distilled water. Analyses were performed under the following conditions:

1. Injection volume: 100 μl
2. Guard column: Brownlee, C_{18}, 5-μ particle size
3. Analytical column: Supelco LC 18, 15 cm × 4.6 mm, 5-μ particle size
4. Mobile phase: 0.05 M $NH_4H_2PO_4/H_3PO_4$, pH 2.5 buffer/CH_3CN
5. Linear gradient profile: 93% buffer/7% CH_3CN; change to 80% buffer/20% CH_3CN in 20 min; change to 65% buffer/35% CH_3CN in 10 min
6. Flow rate: 1.5 ml/min
7. Detector: UV, 280 nm

The HPLC profiles contained about 50 peaks. A typical profile is shown in Figure 1c. The identities were known for eight of the compounds. These were gallic acid, theobromine, theophylline, caffeine, and the catechin flavanol derivatives given in Table 3. Standards were available for gallic acid, theobromine, theophylline, caffeine, and epicatechin. The concentrations of epigallocatechin (EGC), epicatechin gallate (ECG), and epigallocatechin gallate (EGCG) were calculated on a molarity basis by ratio with the epicatechin standard.

Tea quality is often correlated with catechin levels in the fresh leaf. The catechins possess an astringent taste and play the most significant role of any group of substances during the course of tea manufacture. The relative proportion of the various cathechins present affect the course of oxidation and the nature of the final product.[6]

4. Organic Acid Profiles

Low levels of citric and malic acids are present in green tea. In addition, these acids are present in citrus fruits and contribute to the overall citrus impression by providing the characteristic sour note. While black teas are generally not considered to be sour, the possibility exists that this sensation is used to balance the flavor impression of the orange and spice.

The organic acid profiles were obtained by isocratic HPLC. Samples were prepared for analysis by passing them through a C_{18} SPE cartridge. The sample (3 ml) was passed through the preconditioned (methanol/water) cartridge to preload it. The eluate was discarded. A second 3 ml aliquot was then passed through the SPE and the eluate collected for analysis. This approach retains unwanted (interfering) compounds such as polyphenols on the SPE cartridge. The organic acids are not retained on the SPE cartridge under these conditions. Analyses were performed under the following conditions:

1. Injection volume: 50 µl
2. Analytical column: Biorad HPX-87H, 30 cm × 7.8 mm
3. Mobile phase: 0.0015 N H_2SO_4, pH 2.7
4. Flow rate: 0.8 ml/min
5. Detector: UV, 210 nm

Measurable levels were found for citric, malic, succinic, and fumaric acids. Except for fumaric acid, the others are reported to occur naturally in black teas. The levels found are summarized in Table 3. Assuming 100% extraction of the acids from the dried tea solids during brewing, these levels appear to be derived from only the tea.[6]

REFERENCES

1. **Meilgard, M., Civille, G. V., and Carr, B. T.,** *Sensory Evaluation Techniques.* CRC Press, Boca Raton, FL, 1987.
2. **Powers, J. J. and Ware, G. O.,** Discriminant analysis, in *Statistical Procedures in Food Research,* Piggot, J. R., Ed., Elsevier, New York, 1986.
3. **Wu, L. S., Bargmann, R. E., and Powers, J. J.,** Factor analysis applied to wine descriptors, *J. Food Sci.,* 42, 944, 1977.
4. **Piggot, J. R. and Canaway, P. R.,** Finding the words for it — methods and uses of descriptive sensory analysis, in *Flavour,* Schrier, P., Ed., Walter de Gruyter, Berlin, 1981.
5. **Parducci, A.,** Contextual effects: a range-frequency analysis, in *Handbook of Perception,* Vol. 3, Carterette, E. C. and Friedman, M. P., Eds., Academic Press, New York, 1974.
6. **Graham, H. N.,** Tea: the plant and its manufacture; chemistry and consumption of the beverage, in *The Methylxanthine Beverages and Foods: Chemistry, Consumption and Health Effects,* Spiller, G. A., Ed., Alan R. Liss, New York, 1984.

Chapter 9

CORRELATION AND REGRESSION

I. INTRODUCTION

Correlated variables are variables which change together in a systematic fashion, and a correlation coefficient is a measure of the degree and direction of that relationship. Correlations are important because they allow prediction. Correlations were already used in Chapter 6 in the form of prediction equations. In those cases, correlations were determined between changes in concentration of sugar in an orange beverage and the number of observers who noticed increases or decreases in the intensities of various attributes of the products. The correlations were incorporated into regression equations to allow prediction of how similar groups of observers might respond to future changes in the product. Sugar concentration was used as the predictor, or independent, variable, and the number of observers noticing changes in the samples was used as the predicted, or dependent, variable. This chapter will examine the rationale behind these procedures. The mechanics of computing correlations and regression lines is now an easy matter: even the simplest of hand calculators will allow quick computation of correlations and parameters of regression lines. Personal computer plotting software will routinely do these calculations as well. The mechanics are therefore not an issue here, but the conceptual understanding is.

II. CORRELATION

As stated above, correlated variables are variables which change together in a systematic fashion. The most widely used measure of correlation is Pearson's Product Moment Correlation Coefficient, symbolized by r. Although other correlations are available, the one being referred to here will always be Pearson's r. The relationship between correlated variables can be visualized in the form of a *scatterplot*. This is a plot in X and Y coordinates of two different measurements made on each of a series of product samples. One measure is plotted as the Y coordinate, and the other as the X coordinate of a two-dimensional graph. A scatterplot provides considerable information about the relationship between the variables of interest. Figure 1 shows the type of information available. Figure 1A and B show positive and negative correlations, respectively. Both positive and negative correlations provide equally useful information. When the points on a scatterplot tend to rise from the lower left of the plot to the upper right, it indicates a positive correlation. This means that as one variable increases in value so does the other. For example, in Chapter 6 it was shown that the number of observers who will report increased orange flavor in a comparison sample relative to a standard increased as the concentration of sugar in the comparison increased. Therefore, a positive correlation exists between number of observers reporting increased orange flavor and sugar concentration. Figure 1B shows a negative correlation. In negative correlations, the data slope downward from the upper left corner of the graph to the lower right corner. This means that as one variable increases the other decreases. In Chapter 6 it was also shown that the number of sour reports for a comparison sample decreased as sugar concentration of the comparison increased, that is, reports of sourness were negatively correlated with increases in sugar. Correlations range in degree from −1.0 through 0.0 and up to +1.0. Figure 1C shows a perfect +1.0 correlation. When r = +1.0 or −1.0, all of the data points will fall exactly on a straight line. This is a very useful situation because an investigator holding such data will be able to predict *exactly* what the value of one of the variables will be just by knowing the other. For example, with Figure 1C the investigator could predict what value of Y that would be expected

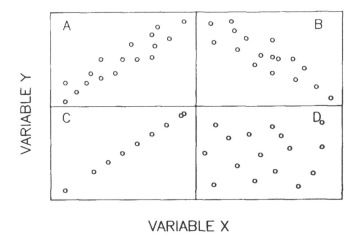

FIGURE 1. Scatter plots representing different direction and magnitude of correlation. (A) positive correlation, (B) negative correlation, (C) perfect positive correlation of 1.0, and (D) zero correlation.

if the X value was equal to 4.3. This would be done by drawing a perpendicular line from the X-axis at 4.3 up to the line connecting the data points and then projecting the line horizontally to the Y-axis. The point at which the horizontal line intersects the Y-axis corresponds to the predicted Y value. Alternatively, the formula for the straight line connecting the points could be found and any value of Y corresponding to any value of X could be found by appropriately solving the equation, as was done with thresholds in Chapter 6 (even though the correlation was less than 1.0 in that case).

Correlations of +1.0 or –1.0 are virtually never found. Usually an investigator will be very happy to see data plotted as in Figure 1A and B which indicate reasonably good correlations. In general, the closer the data points seem to cluster toward an imaginary line through the center of the pattern of points, the higher the correlation will be and prediction will be more precise. Figure 1D shows how a 0.0 correlation would look in a scatterplot, the lack of discernible trend in this figure is what indicates 0.0 correlation. Numerically, both positive and negative correlation coefficients of around 0.4 to 0.6 can be considered as of moderate size. Correlations greater than 0.6, of either sign, are very good, and those below 0.3 are less useful.

A. INTERPRETING CORRELATIONS

A correlation between two variables does not imply a cause-effect relationship. It only means that the two variables change in a systematic fashion relative to one another. For example, the concentration of a certain chemical compound in a food product may be found to be correlated with acceptability of the product. This does not necessarily mean that the compound is responsible for acceptability. It may be, for example, that increases in the concentration of the compound only serve to mask the presence of some other undesirable flavor which is, in fact, controlling acceptability. As concentration increases, masking of the undesirable flavor becomes greater and acceptability therefore increases. In that case, the compound in question is only spuriously related to acceptability by virtue of its ability to mask a negative which is, in fact controlling acceptance.

Correlations are independent of the units in which measurements are taken. In fact the measurements themselves need not even be in the same units in order to obtain data and compute correlations. For example, molar concentration of a chemical may be correlated with measurements of sensory intensity obtained by magnitude estimation, viscosity in centipoise, peak height of a chromatogram, etc. Further, any of these measures (or any others) may all be correlated with each other for useful information. Figure 2 shows an example of how cor-

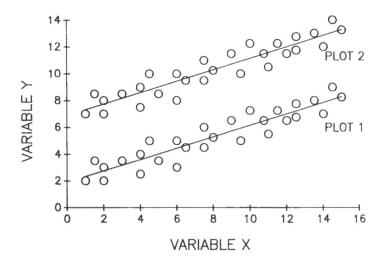

FIGURE 2. The effects of adding a constant to each data point of one of two variables being correlated. In this case a value of 5 was added to each of the original values of Y used in the correlation shown in plot 1. The resulting plot 2 has the same correlation between X and Y, 0.92, as does plot 1.

relations are independent of units of measurement. The scatterplot labeled plot 1 is constructed from a set of hypothetical data points which result in a correlation of 0.92 between the values of X and Y. Plot 2 is the resulting scatterplot when a value of 5 is added to each of the Y values shown in plot 1. The correlation between the X values and the new Y values in plot 2 remains 0.92. Both correlations have exactly the same meaning and usefulness in prediction. The only requirement of measures used to compute Pearson's r is that they be linearly related.

Constricting, or shortening, the range of either the X or Y variable, or both, will greatly reduce the size of the obtained correlation. This is shown in Figure 3. When the full range of values shown in this plot is used to compute a correlation, the result is 0.85. This is a substantial relationship. However, if only the values enclosed by the box in the lower left of the figure are used for computing the correlation, a value of 0.24 is obtained. This value is of little, if any, use in prediction. This illustrates the importance of including a substantial range of each variable in chemometric studies if existing relationships are to be found. If product samples are restricted to only those which closely approximate what is likely to be the final execution, then variable ranges will be artificially restricted and important sensory-physical relationships will not be found. Consequently, the investigator will not obtain knowledge of how composition changes will affect product perception.

Perhaps the most useful piece of information obtained from a correlation is the *coefficient of determination*. This is simply the square of the correlation coefficient. For example, the coefficient of determination for a correlation of 0.9 is equal to 0.81. This means that 81% of the changes observed in one of the variables are directly accountable for by changes in the other variable. Conversely, only 19% of the variability in one of the variables *cannot* be accounted for by changes in the other variable. The coefficient of determination, or r^2, provides an index of how useful a correlation will be. Thus, while a correlation of 0.7 is a good correlation, it can only be used to predict 49% of the variability that is observed. Over half of the observed differences in the measure of interest cannot be predicted by knowledge of the other variable.

III. REGRESSION

Regression and correlation are virtually interchangeable terms in statistics. Regression

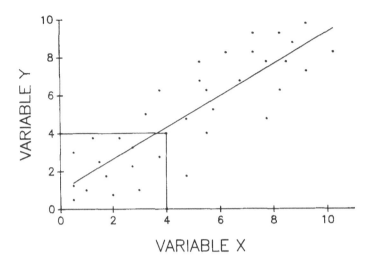

FIGURE 3. Effects of restricting the range of variables on the magnitude of the correlation coefficient. The correlation obtained when the full range of values shown on the graph is used is 0.85. When only the points included in the box in the lower left corner of the graph are used the correlation drops to 0.24.

analysis is necessitated by the fact that correlations are virtually always less than +1.0 or −1.0. It was noted above that if points of a scatterplot all fell exactly on a straight line, then predicting results of a future study would be an easy matter of reading the graph. The typical situation is more like that shown in Figure 1A and B. In those cases, a straight line can be drawn through the center of the pattern of points, but it will not touch all, or even most, of them. In fact, it will touch very few. Regression provides a way of dealing with this situation which results in the best predictions possible from data that are less than perfectly correlated.

Measurement values of any given phenomenon tend to differ from one another but cluster around an average value. Regression is the term used to describe the clustering toward an average value. For example, chemical determinations of tannins in tea, will vary from sample to sample of the same tea but will remain more or less around an average value for that tea. Likewise, sensory evaluation ratings of the astringency of the tea will vary from sample to sample and from one observer to another, but they too will cluster about an average value. The sample-to-sample and rating-to-rating variability is error variability. The idea of the regression line is to estimate where the true values of the ratings or chemical determinations actually lie and then to draw a line through those points. Predictions are then made from this regression line with the knowledge that predicted values will only be somewhere near the true value because of the error variability inherent in any measurement. This is what was done in Chapter 6 when approximate thresholds were estimated from straight lines drawn through the obtained data.

Regression lines are determined by using the *least squares* technique to result in what is called the *least squares* or *best fit* regression line. Computation of these lines can be done on almost all hand calculators and personal computers, and the mechanics are described in numerous sources.[1-5] The instruction handbook for most hand calculators will also describe the procedure. The logic of the procedure is as follows. It is assumed that the best line to make predictions from is the one which comes as close as possible to all the obtained data points; that is, the sum of the distances between the regression line and each data point should be as small as possible. In practice this means that there will be an equal number of points above and below the line, and the sum of the distances between the line and all points above it will be equal to the sum of the distances between the line and all points below it. This situation presents a mathematical dilemma. That is, equalizing distances between the line and points above it and below it means that the sum of *all* distances will be zero because positive values will exactly balance negative values. This problem is solved by squaring the distances before summing them. The objective is then to minimize the value of the *squared* distances between the line and all data points. That is why the line is referred to as the *least squares regression* line. The word regression is used to describe the line because it is assumed that with repeated determinations of the data points there will be a tendency for them to cluster, or regress, toward the line. The line is therefore assumed to be an estimate of what the average values would turn out to be in the long run.

A. SELECTING A REGRESSION LINE

Most computer computational routines for regression analysis can be executed in a way that allows all variables in a study, as well as their squares and interaction terms, to be included in a regression equation. In practice it is usually found that only a few of the possible components of a regression model (i.e., equation) will be important in prediction. It is therefore often valuable to include in the analysis only those variables likely to be important. This is so because blind analysis by a computer can yield spurious results. For example, two variables may be correlated because of their correlation with a third variable. Even though the two variables may have no direct relationship with each other, a computer routine may find them as predictive of the phenomenon in question because of their relationship to the third variable which is truly important to prediction. An investigator with knowledge of the material being studied will be able to specify which variables are likely to contribute most heavily to accurate prediction and direct the computer program to use only those variables. This can usually be done by visual inspection of plots of the data and is discussed in this section of the chapter.

1. First Order Regression

A first order regression equation is the equation for a straight line. A large number of chemometric applications can be done using first order regression. The example used here will be taken from the threshold data discussed in Chapter 6. Among other things, these data specified the proportion of times each of a series of comparison beverages of increasing sugar concentration was judged to be more sweet than a standard beverage of constant sugar concentration. When the proportion of "more sweet" responses was plotted as a function of concentration of sugar in the comparison, a very smooth S-shaped, or ogive, curve was obtained. At this point a regression analysis which determined the formula of the line best describing the ogive could have been performed. However, it would be much easier if a simpler regression line could be found. As it turns out, anytime data of this sort plot as an S-shaped curve, a transformation of the data points to z-scores will convert the plot to a straight line. This was done in Chapter 6 and the figure that resulted is reproduced here (Figure 4).

It can be seen from this figure, that after conversion to z-scores, a straight line can be drawn through the data points and come reasonably close to all of them. Whenever this is the case, a *first order regression* is specified. This is another way of saying that the data can probably be described by a straight line, and therefore the same straight line will likely allow good prediction of future results. The general equation for a straight line is

$$Y = mX + b \tag{1}$$

where Y = the values on the ordinate which are the ones that are being predicted or described, m = the slope of the line which indicates how much the value of Y will change with each change in the value of the predictor variable, X, and b is the projected point at which the straight line will intersect the Y axis. The straight line that was found to describe the data in Figure 4 is

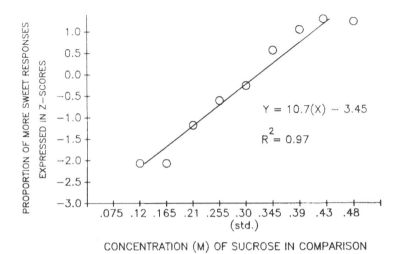

FIGURE 4. Proportion of "more sweet" responses given to each of a series of sucrose concentrations compared to a standard 0.30 *M* solution. The straight line is the least squares regression line fitted to the data such that it comes as close as possible to each point. The R^2 of 0.97 indicates that the correlation between the values actually obtained in the study and those predicted from the regression line is high.

$$Y = 10.7(X) - 3.45 \qquad (2)$$

This equation means that to predict what value of Y is expected for a given value of X, the X value should be multiplied by 10.7 and then subtract 3.45 from that product. If the value of X were 0, then the predicted value of Y would be –3.45. A z-score of –3.45 is nearly three and one half standard deviations below the mean of a normal distribution. Therefore, this equation predicts that if there were no sugar in the beverage from which these data were derived, virtually everyone would say that such a sample is less sweet than the standard of 0.3 *M* sugar concentration. This certainly makes intuitive sense.

The slope of 10.7 may at first glance seem to contradict the visual impression from the graph itself where the slope seems close to 1.0. The apparent contradiction is because the sugar concentration units on the X-axis are in the range of one tenth of the size of the z-score units on the Y-axis. The graph also shows that the regression line comes very close to the actual data points. This implies that the correlation between the obtained data points and the points predicted by the regression line will be very high. In fact, that correlation, or R^2, is 0.97, suggesting that this equation will be very useful in predicting how noticeable changes in sugar concentration will be among the population in future studies of new formulations. R^2 is the coefficient of determination applied to regression analysis and its interpretation is the same as that for r^2 discussed with correlation in general.

2. Second Order Regression

Straight lines cannot always describe obtained data. Figure 5, also taken from the threshold study in Chapter 2, is a plot of the proportion of more orange flavor responses to a series of samples judged against a standard as a function of increasing sugar concentration in the comparison. While a straight line could certainly be drawn through these points, the line would come closer to more points if it had a slight downward curve to it as it does in Figure 5. Whenever data plot with a downward curvature (or negative acceleration) *or* and upward curvature (positive acceleration), a *second order regression* equation will describe the data. The second order regression is distinguished by the appearance of a squared, or quadratic, term in the formula. The general form of the quadratic equation is

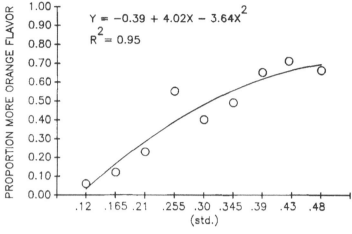

FIGURE 5. Proportion of "more orange flavor" responses given to each of a series of orange beverage samples compared to a standard sample. The actual orange flavor level was constant, but perceived orange flavor increased with increases in the sugar concentration of the samples as compared to the standard 0.30 M sucrose sample. Increases in the proportion of "more orange flavor" responses is a concave downward, or negatively accelerated, function of sugar concentration. This type of function is described by a quadratic equation which accounts for the slower rate of increase in the values of Y as values of X increase.

$$Y = m_1(x) + m_2(x^2) + b \qquad (3)$$

The quadratic term accounts for the curvature of the line. In relative terms, the coefficient (i.e., "m") of the squared term will be small. As a consequence, small values of X will not have a large effect on the solution of the equation. However, as the value of X increases, its impact on the equation will quickly increase because the X value is squared before being multiplied by its coefficient. Thus, if the curvature of the line is downward, the quadratic term will subtract more and more from the value of Y as the value of X increases. This situation is exemplified in Table 1 using the equation which describes the data in Figure 5:

$$Y = -0.39 + 4.02(X) - 3.64(X^2) \qquad (4)$$

It can be seen that if the value of X is increased sufficiently, the value of Y would actually start to decrease. Although this is a dangerous extrapolation from the actually obtained data, it is reasonable to suppose that perceived orange flavor would be increasingly masked if sweetness were great enough, and the investigator might wish to follow up that hypothesis with an additional study.

The coefficients in the quadratic equation are interpreted in a fashion similar to those in the first order equation, that is, the first order, or linear, component describes the extent to which Y increases with increases in X; the squared, or quadratic, component describes the extent to which higher values of X subtract (or add) to Y at higher values of X, and the intercept is the extrapolated value of Y when X is zero. R^2 is still the correlation between the predicted and obtained data points. In the example being used, $R^2 = 0.94$, which is quite good.

Data that have a negative or positive acceleration when plotted in linear coordinates can be described by a straight line if the numbers are first converted to logs. If that is done, a first order regression line can be fit to the logs and interpreted in the usual fashion. However, if

TABLE 1
The Effects of Increasing Values of a Predictor Variable, X, on Values of the Predicted Y Variable in a Second Order Regression Equation

$$Y = -0.39 + 4.02(X) - 3.64(X^2)$$

X Value	Predicted Y	Subtraction due to quadratic term	Successive increments in value of Y
0.12	0.04	0.05	—
0.165	0.18	0.09	0.14
0.21	0.29	0.16	0.11
0.255	0.40	0.23	0.11
0.30	0.49	0.32	0.09
0.345	0.56	0.43	0.07
0.39	0.62	0.55	0.06
0.43	0.66	0.67	0.04
0.48	0.70	0.83	0.04

the curvature of the original line in linear coordinates is such that the value of Y actually begins to change direction and decrease at high values of X, then a log transformation and a first order regression is not appropriate because it cannot account for the change in direction. The value of the quadratic equation is that it can account for such a change in direction and ought to be used when such a change is found. For example, it is well known that acceptability of many products increases as attribute intensities increase, but then decreases when intensities become high enough.[6] Thus, acceptability data ought to generally be described by quadratic equations.

3. Interactions between Variables

The statistical definition of interaction is the change in the effects of one variable as a function of a change in a second variable. When this occurs, the two variables are said to interact. In Chapter 7 the use of factorial designs in collecting sensory data was discussed. Factorial designs are used when an investigator wishes to determine if or how two or more variables interact to determine the sensory properties of the products being studied. To do that, a set of products is formulated such that each level of one of the variables of interest is combined with each level of the the other variable of interest. The study discussed in Chapter 7 combined five concentrations of sucrose with three concentrations of citric acid to determine how, among other things, sucrose and citric acid would interact to produce the perceived level of sweetness. Table 2 shows the design of the study schematically and Figure 6 shows the average sweetness ratings from the 30 observers who tasted and rated each product on a 20-point category rating scale of sweetness intensity. This graph shows several pieces of information that have already been discussed as useful for selecting a regression equation and some new information as well. First, the graph consists of three lines, each representing the sweetness ratings of successive concentrations of sucrose at one of the three citric acid concentrations. Each line shows the same general trend of increasing in height with increases in sucrose concentration as is expected. Consequently, it can be concluded that the regression model will contain a linear (first order) component for sucrose. Second, as citric acid *increases* in concentration, sweetness ratings *decrease*, and this effect becomes more pronounced at higher levels of citric acid. As a consequence, each of the three lines in the graph has a progressively shallower slope, indicating that we should expect an interaction between citric acid and sucrose, that is, the effect of citric acid will depend on which level of sucrose it is combined with. Each of the three lines also appears somewhat negatively accelerated, so a

TABLE 2
Factorial Design Used to Determine the Effects of Five Levels of Sweetener and Three Levels of Citric Acid on the Perceived Orange Flavor Intensity of an Orange Beverage

Citric acid concentration (M)	Sucrose concentration (M)				
	0.12	0.21	0.30	0.39	0.48
0.003	0.12/0.003	0.21/0.003	0.30/0.003	0.39/0.003	0.48/0.003
0.0125	0.12/0.0125	0.21/0.0125	0.30/0.0125	0.39/0.0125	0.48/0.0125
0.05	0.12/0.05	0.21/0.05	0.30/0.05	0.39/0.05	0.48/0.05

Note: A total of 15 beverages was prepared. Each of the beverages contained one of the five levels of sucrose listed across the top of the table. At each of the five sucrose levels, one product sample contained 0.003 M citric acid, one contained 0.0125 M citric acid, and one contained 0.05 M citric acid. Therefore, each level of sucrose was mixed with each level of citric acid to allow determination of how the two variables interact. Columns in the table represent sucrose concentration and rows represent citric acid concentration. Each cell represents one of the 15 treatments.

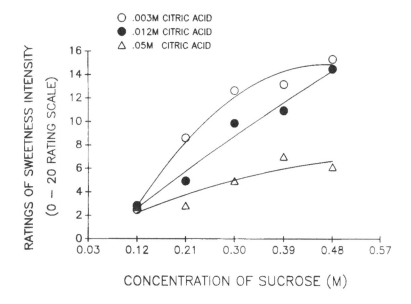

FIGURE 6. Example of data which requires an interaction term in the regression equation for adequate description. The interaction term is required because each of the three lines has a different slope. Therefore, the effects of one variable depend upon which level of the other variable they are combined with. In this case, sucrose behaves differently at each of the three levels of citric acid it is combined with.

quadratic component for sucrose is likely to be needed. Finally, although not clear in the figure, it has been reported that citric acid has a slight sweet taste in addition to its stronger sour taste.[7] Therefore, a small positive weight for a linear citric acid component in the equation may be expected.

In summary, visual inspection of Figure 6 suggests that a regression equation to describe the data it contains might include at least a first order component for both sucrose and citric acid, a second order component for sucrose, and finally a term to account for the apparent interaction between citric acid and sucrose. The general model suggested by the figure is

$$Y = m_1 (X_s) + m_2(X_c) + m_3(Xs^2) + m_4(X_sX_c) + b \tag{5}$$

In this model, X_s = sucrose concentration and X_c = citric acid concentration. When stating general models as is done above, it is customary to designate all weights (i.e., m) as being positive. The actual analysis will specify both the size and direction of the weights. This model states that the sweetness ratings will be predictable from the concentration of sucrose, citric acid, the squared value of sucrose, and the interaction of sucrose and citric acid. The interaction is stated as X_sX_c and means that the effect of sucrose on the ratings will depend on which level of citric acid it is mixed with. An actual regression analysis provided the following equation based on the data from Figure 6:

$$Y = 62.06(X_s) + 27.13(X_c) - 42.47(X_s^2) - 479.28(X_sX_c) - 3.63 \tag{6}$$

The final model therefore says that, in this study, sweetness ratings were determined by the linear effects of sucrose and citric acid concentration, the interaction of these two variables, and the quadratic component from sucrose concentration. Study of the relative size of the weights applied to the terms in the equation reveals how strongly each of the variables is involved in determining the sweetness ratings. For example, the weight for the linear sucrose component is 62.06. This is relatively large when the concentrations of sucrose are considered. If 62.06 is multiplied by 0.48 *M* sucrose, the value is found to be 29.78. This means that 29.78 rating points would be added to sweetness ratings by that concentration of sucrose. On the other hand, the weight for the linear component of citric acid is 27.13. The concentrations of citric acid range only up to 0.05 *M*. Multiplying 0.05 *M* citric acid by 27.13 results in a value of 1.35. Consequently, only 1.35 ratings points for sweetness would be added by 0.05 *M* citric acid. This makes sense because citric acid does have a slight sweet taste,[7] but is much more sour than it is sweet. The sourness ought to suppress sweetness as concentration increases. This is reflected in the large value of –479.28 associated with the interaction of sucrose and citric acid. Although citric acid may add a bit of sweetness, its ability to suppress sweetness increases dramatically as its concentration increases. Thus, the negative effect of citric acid on sweetness ratings appears in the interaction term, reflecting the greater ability of citric acid to suppress sweetness when at a higher concentration. This regression therefore separated the strong, sweet-suppressing effect of citric acid from its very slight sweet taste component. Since the linear effect of citric acid is in fact so small, the investigator may wish to leave it out of the equation altogether. In this case it was decided to retain the term in the equation because of the known slight sweetness of citric acid, and also because it seems intuitively reasonable to include the individual components of significant interactions in the model. However, the term could be eliminated with little loss of predictive power.

IV. DETERMINING EQUATIONS STATISTICALLY

The preceding section discussed how to hypothesize about what sort of regression equations would likely fit a set of data on the basis of how the data appeared in a scatterplot. With a little practice this can become quite easy, especially when only a small number of variables is involved and the investigator is familiar with how those variables generally behave. When more than a few variables are involved in a study, the task becomes very difficult. In those cases, *stepwise regression* procedures can be used. Stepwise regression is the term applied to the process of selecting the best regression equation by iteratively computing all possible equations and selecting the one that produces the best description of the data. The procedure is done by computer and routines for carrying it out are widely available.

This is how the procedure works. Suppose an investigator has collected flavor intensity estimates of a given set of samples and has determined the concentrations of five chemicals in the samples which are believed to influence flavor intensity. The objective is to predict flavor intensity from knowledge of concentrations of the five chemicals. In the first step of the stepwise regression, five separate linear, or first order, regressions are computed, i.e., one for each of the five chemicals. Each of these equations will indicate how well flavor intensity can be predicted from one of the chemicals. A determination is then made of which chemical provides the best prediction. This is done by comparing values of R^2, with the largest value being selected as indicating the best single predictor variable. Recall that R^2 is the correlation between the actual obtained values (in this case, ratings of sensory intensity) and the values predicted by the regression equation. Consequently, the largest R^2 value will be associated with the single variable that allows the regression line to come the closest to all the obtained data points.

After the first predictor variable is selected as described above, the procedure then adds each of the other variables to the equation, one at a time, and computes the R^2 value for each of the the equations with two predictor variables included. The variable which, when added to the equation, results in the largest increment of R^2 is included. This process is repeated for each combination of three, four, and finally the five variables. Then the procedure begins to add the squared, or quadratic, values of each variable in a similar fashion. This is carried on until each combination of the variables, their squared terms, and any interactions is added to the regression equation and R^2 values are computed.

The selection of the final equation is not done blindly by the computer. Instead, the printout of the procedure is a listing of the equation and associated R^2 value obtained at each step. The investigator then inspects this list and selects the equation that provides the most useful information. This is done by inspection of the R^2 values. In general, as successive terms are added to the equation, the value of R^2 will increase by successively smaller amounts. The addition of terms to the equation which no longer sufficiently increase the value of R^2 indicates that those terms need not be included in the equation selected for prediction, that is, at some point the investigator decides that the increase in predictive power gained by an additional term is not large enough to warrant inclusion of that term in the equation.

Table 3 is an example of the output of a stepwise regression. The analysis was performed on the data used in the example of predicting sweetness of an orange beverage from the concentrations of sucrose and citric acid in the formulations. These are the data plotted in Figure 6. In this example, the R^2 values give an indication of how much predictive power is gained from each term and combination of terms added to the equation. For example, neither sucrose nor citric acid concentration alone describe the data very well, as indicated by the modest R^2 values associated with them. However, when the linear sucrose component and the linear citric acid component are both included in the equation, the R^2 value jumps to 0.86. This indicates that both terms ought to be included in the final model. The next step of the selection process is to determine if the addition of a third term to the equation adds more explanatory power. Table 3 shows that adding the quadratic term for citric acid to the equation does not result in an increase in R^2 and the quadratic sucrose term only results in an increase up to 0.87. When the interaction of sucrose and citric acid are considered, the value of R^2 jumps to 0.93, indicating that this interaction plays an important part in determining the sweetness ratings. The next section of Table 3 shows the results of regressions done adding two additional terms to the original sucrose and citric acid linear terms in the model. Again, adding quadratic terms for both sucrose and citric acid result in only a modest gain in predictive power. When the interaction term (sucrose × citric acid) is added along with either the quadratic citric acid term *or* the quadratic sucrose term, the value of R^2 reaches 0.94 and 0.95, respectively. This value is not further improved by an equation which includes all five possible terms. Consequently, the linear sucrose and citric acid terms, the quadratic sucrose term, and the interaction term are the ones selected for the final model because the combination of those terms results in the highest value of R^2.

TABLE 3
Increases in the Value of R² with the
Stepwise Addition of Predictor Variables to
the Regression Equation

Predictor variables in equation [a]	R^2
SUC	0.56
CIT	0.28
SUC, CIT	0.86
SUC, CIT, CIT²	0.86
SUC, CIT, SUC²	0.87
SUC, CIT, SUC × CIT	0.93
SUC, CIT, SUC², CIT²	0.87
SUC, CIT, CIT², SUC × CIT	0.94
SUC, CIT, SUC², SUC × CIT	0.95
SUC, CIT, SUC², CIT², SUC × CIT	0.95

[a] SUC = sucrose, CIT = citric acid.

Table 4 lists the results obtained when the regression equation just selected to describe the data shown in Figure 6 is solved for each of the sucrose and citric acid concentrations used in the original study. The first two columns in the table list the values of sucrose and citric acid substituted in the equation for solution. The third column lists the actual sweetness ratings obtained for the combination of sucrose and citric acid listed in the first two columns. The fourth column shows the sweetness rating predicted by solution of the regression equation derived from the data. Finally, the fifth column lists the differences between the actually obtained ratings and the predicted ratings obtained from solving the equation for each set of sucrose and citric acid values. These *residuals* are all very small and relatively constant in size, indicating that the obtained equation makes accurate predictions of sweetness ratings across the entire range of samples used in the study.

V. ISSUES IN REGRESSION ANALYSIS

The central question in regression analysis is how to select the final equation to describe the obtained data and predict future results. If only one dependent variable and one predictor variable is used, the answer to this question is simple because the possible regression equations are simple. As the number of possible predictor variables increases, the number and complexity of alternative regression equations to describe the data increases very quickly and makes decision more difficult. No hard and fast rules exist, but there are a number of criteria that can be employed to facilitate selection.

The most direct criterion for equation selection is the value of R^2. This value should be selected to be as large as possible while maintaining the simplest equation. If the addition of another term to an equation does not serve to usefully increase R^2, then that term should not be included. The size of residuals also provide guidance. The differences between actually obtained values and predicted values ought to be consistently small. For example, if the residuals at both extremes of a straight line regression are larger than the residuals calculated for the middle of the line, it could indicate that a curved line would provide a better description of the data. In that case, the investigator would include a quadratic term in the equation to account for the curvature. Most computer programs will provide a statistical test of each term included in the equation which indicates whether or not its contribution to predictive value is significant. Finally, knowledge of how the chemical or sensory system being studied works should enter into judgment of whether or not a term should be added to an equation.

Another important issue in regression is the problem of *multicollinearity*. This problem

TABLE 4
Obtained Sweetness Ratings and Ratings Predicted by Regression Analysis

Regression equation:

$$Y = 62.06(X_s) + 27.13(X_c) - 479.28(X_s x X_c) - 42.47(X_s^2) - 3.63$$

| Molar conc | | | | |
SUC	CIT	Obtained value	Predicted value	Residual
0.12	0.003	2.48	3.10	−0.62
0.21	0.003	8.61	7.30	1.31
0.30	0.003	12.64	10.81	1.83
0.39	0.003	13.16	13.63	−0.47
0.48	0.003	15.32	15.76	−0.44
0.12	0.0125	2.83	2.82	0.01
0.21	0.0125	4.93	6.60	−1.67
0.30	0.0125	9.87	9.70	0.17
0.39	0.0125	10.93	12.11	−1.18
0.48	0.0125	14.48	13.83	0.65
0.12	0.05	2.70	1.68	1.02
0.21	0.05	2.83	3.84	−1.01
0.30	0.05	4.93	5.32	−0.39
0.39	0.05	7.00	6.12	0.88
0.48	0.05	6.12	6.22	−0.10

occurs when two or more of the predictor variables included in a regression equation are highly correlated with one another. When this happens, it is impossible to determine which of the variables is actually predictive of data being considered. An excellent example of this is given by Schroeder et al.[4] using traffic fatalities as the variable being predicted. In the state of New York, mandatory seat belt laws were imposed at about the same time enforcement of drunk driving laws was greatly increased. A subsequent decrease in traffic fatalities could therefore not be exclusively attributed to either of the variables. In chemometrics, whether chemical or sensory data are considered, predictor variables are very frequently correlated to some degree and often nothing can be done about it, especially if the investigator lacks other evidence to suggest which of the variables is truly responsible for variation in the predicted variable. As a general rule, variables included in a regression equation ought to be correlated with one another as little as possible, and correlated with the predicted variable as much as possible. The addition of a predictor which is correlated with another predictor already in the equation will add relatively little to the value of R^2. This is because the added predictor variable is predicting much of the same variability that the other predictor variable it is correlated with is accounting for. On the other hand, two predictors which are both correlated with the predicted variable but not with each other will add greatly to the value of R^2 because each is predicting variability not accounted for by the other.

Regression analyses assume that the error variability around the regression line is constant over the range of the regression, that is, the size of the residuals is about the same regardless of which part of the line they are calculated from. This is called the assumption of *homoskedasticity*. If the size of the residuals varies in some nonrandom fashion along the range of the regression, then the data are *heteroskedastic* and the regression parameters will be less accurate and prediction less precise. One common cause of heteroskedasticity in data is the fact that means and variances are often positively correlated, that is, variability is greater around larger numbers than it is around smaller numbers. Another cause of heteroskedasticity is the effect of a variable which was not included in the equation but interacts with the ones that were included. If this can be determined to be the case, then inclusion of that variable in the equation, if possible, will eliminate the problem. If it is not the case, or it cannot be determined if it is the case, then a generalized least squares (GLS) analysis may be indicated. These procedures are routinely available in computer statistical packages and present no special executional problems. Basically, a GLS analysis provides differential weighting to the obtained data points. Those points having higher variability are given less weight in the analysis than those having smaller variability. While statistical tests can be performed on the data to determine if heteroskedasticity exists, its detection can be more straightforward. Heteroskedasticity should be suspected if the cloud of data points around a plotted regression line is consistently wider in some areas than others.

A difficulty that can exist when sensory measurements are taken over a series of samples by the same group of observers is called *autocorrelation*. This means that successive data points are correlated, i.e., the ratings given to any given sample are determined in part by the ratings given to the immediately preceding sample. The presence of autocorrelation results in decreased reliability of the estimated regression parameters and decreased accuracy of prediction. This is because earlier ratings have acted as an uncontrolled variable in determining later ratings and that variable is not included in the regression equation. This effect can be handled to some extent by ensuring that each observer is presented with a different random order of product samples, or that the different orders that can occur do so equally often across the group of observers. This will not eliminate the dependence of ratings on prior ratings, but it will balance the effects of autocorrelation evenly across all products. In time-intensity studies where the same attribute is rated several times across a period of many minutes, the effects of autocorrelation cannot be eliminated by experimental design manipulations. Autocorrelation is detected by examination of residuals. If residuals of predicted ratings on trial t are correlated with residuals for predicted ratings on trial $t - 1$, then autocorrelation is indicated. Again, the GLS procedures discussed in association with heteroskedasticity will allow analyses to be performed with accurate results and prediction by differentially weighting the residuals when the least squares regression line is determined.

REFERENCES

1. **Edwards, A. L.,** *Multiple Regression and the Analysis of Variance and Covariance*, W. H. Freeman, New York, 1985.
2. **Merrill, A.,** *Fitting Linear and Curvilinear Regression Lines with a Pocket Calculator*, Merrill Analysis, New York, 1978.
3. **Lewis-Beck, M. S.,** *Applied Regression: An Introduction*, Sage, Beverly Hills, CA, 1989.
4. **Schroeder, L. D., Sjoquist, D. L., and Stephan, P. E.,** *Understanding Regression Analysis: An Introductory Guide*. Sage, Beverly Hills, 1986.
5. **Berry, W. D. and Feldman, S.,** *Multiple Regression in Practice*, Sage, Newbury Park, 1985.
6. **Pangborn, R.,** Individuality in responses to sensory stimuli, Solms, J. and Hall, R. L., Eds., *Criteria of Food Acceptance: How Man Chooses What He Eats*, Forster-Verlag, Zurich, 1981.
7. **McBurney, D. H. and Bartoshuk, L. M.,** Interactions between stimuli with different taste qualities, *Physiol. Behav.*, 10, 1101, 1973.

Chapter 10

DISCRIMINANT ANALYSIS

I. SIMILARITIES AND DIFFERENCES IN SAMPLES

The simplest level of comparison between two products is to search for and identify where similarities and differences exist in the data that describe them. This is what we do intuitively in making many everyday decisions. The process can be summarized in two questions. What does product A have that product B does not, and vice versa? What is the effect of the similarities and differences on product performance?

Differences between products may or may not be significant to the performance of the product. For example, when buying a new automobile, the data gathered might be the manufacturer, body style, engine size, type of transmission, and the color. Two automobiles that are identical, but different in color, will perform the same. The difference between the autos is only important if there is a preference for one color over the other. Two automobiles made by different manufacturers, with different engines and transmissions will probably perform differently, even if they are the same color. The real *performance* difference is in the type of the automobile, the engines, transmissions, and the body style. Color will probably not affect the performance of the vehicle. Nevertheless, color is a variable that was chosen to characterize the automobiles. In analyzing the data gathered for the automobiles, we first search for differences in the characteristics and then determine if these differences are significant with respect to the performance of the automobiles.

The comparison of similarities and differences in food and beverage samples is analogous to the automobile example. Significant differences in sensory perception are meaningful for the comparison of food and beverage samples. Differences in chemical composition may or may not be related to observed sensory differences and may not even be important to the sensory performance. The significance of any chemical changes that are found must be further evaluated. However, before differences can be evaluated for impact, they must be identified.

For chemical and sensory data, the comparison of samples is easy to perform when only a few large differences exist between different types of samples. This can often be done by visual inspection of data tables or graphical displays. Visual inspection of tables of numbers is usually univariate in nature and can only pick out the most obvious changes between samples. When replicate analyses of multiple samples are used to estimate the true value of a variable, manual data interpretation becomes difficult due to the sheer volume of information that must be digested. The level of complexity is further increased when the differences are subtle and involve simultaneous changes in several variables.

Discriminant analysis or *pattern recognition* techniques can be used to identify the similarities and differences in different types of samples providing sample-sample comparisons. The selected data can then be used to predict the characteristics of unknown samples. These techniques operate on the assumptions that samples of the same type are similar, that differences exist between different types of samples, and that these similarities and differences are reflected in the measurements used to characterize the samples.[1] Because they use the *a priori* knowledge of which samples are similar and which are different, discriminant analysis techniques are said to be *supervised* in the development of the discrimination criteria. Techniques that model continuous variations in variables, such as regression or principal components analysis, are *unsupervised* since they do not use information about sample type in the development of the mathematical model.

Discriminant analysis techniques are categorized according to the underlying assumptions used in the development of the models. *Parametric* techniques assume multivariate normal

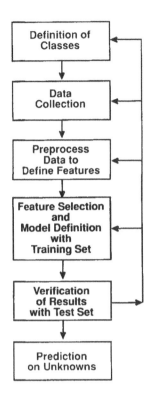

FIGURE 1. Discriminant analysis sequence. The process often requires several iterations before success is obtained. When poor classification results are obtained during feature selection or verification, successive iterations may include (1) redefinition of classes, (2) reevaluation of the transformations used to convert the raw data into features, and (3) collection of additional data if (1) and (2) are not successful.

distributions for the variables. All the statistical requirements for the number of *degrees of freedom* (i.e., the sample to variable ratio) and the homogeneity of the covariance-variance (CV) matrix must be met.[2] It is often difficult to satisfy all the requirements for the application of parametric techniques to chemical and sensory data. *Nonparametric* techniques are not based on assumptions about the data distributions; they let the data define its own structure and are capable of finding that structure either explicitly or implicitly.

II. DISCRIMINANT ANALYSIS STRATEGY

Discriminant analysis techniques can be applied to both chemical and sensory data. For both types of data, the goal of a discriminant analysis is to be able to place a given sample in the group to which it is most similar. The decision about a group membership of a sample is based on the data that were gathered to characterize the samples and the mathematical model that is developed in the process of discriminant analysis.

Discriminant analyses consist of multiple steps as outlined in Figure 1. The specific questions to be answered and the data that have been gathered determine which discriminant analysis technique is most appropriate. Massart et al. summarize general strategies for the selection and application of various discriminant analysis techniques.[2]

A. DEFINITION OF CLASSES

As a part of definition of the data set, attempts must be made to accurately define the number of different types of samples or *classes* present in the data and the membership in the different classes. Proper class definition is crucial to the successful application of discriminant analysis techniques, as the decision rules developed are based on the user-defined classes. If the discriminant analysis techniques are not successful in separating the samples according to the defined classes, either the classes were not defined properly or the data does not contain the necessary information.

Sample classes must be *homogenous* and potentially meaningful. To be homogenous means that the samples assigned to the same class must exhibit some common characteristics and be different from those assigned to a another class. Discriminant analysis techniques will not be particularly successful if one class (e.g., the product of interest) is homogeneous and the second is *nonhomogeneous* or *asymmetric* since it includes "all other samples".[1] This is because "all other samples" may or may not be similar and related to each other. In such a case, more than two classes would be needed to properly define the data set. Underestimation of the actual number of classes can result in overly optimistic estimates of discrimination ability, while overestimation may yield low classification rates. Studying the data interactively, using numerical analysis and graphical displays, will help one to identify outliers and redefine classes if necessary.

There is no all-encompassing set of rules to define the class that a given sample belongs to. A given sample can belong to different classes, depending on its relationship to the other samples in a data set and the questions that are to be answered. Obviously, replicate analyses of the same sample will be assigned to the same class. For most studies, the definition of sample classes is straightforward. The common approach to class definition is based on knowledge of sample origin or history. Samples from a common source, or that received the same processing sequence, are assigned to the same class for comparisons of the effect of sample origin (e.g., supplier, manufacturer, or raw material lot) or for evaluation of the effects of process changes on sample composition. Samples from a second source may or may not be similar to those from the first source and should be assigned to a separate class initially.

For example, tea samples from the same manufacturer, even if they are derived from different lots of raw materials, would be considered to be members of the same class when being compared to products from other manufacturers. The definition of classes by manufacturer allows the comparison (both chemical and sensory) of products in terms of starting materials and processing. On the other hand, to check for product consistency by comparison of the lot-to-lot variation in products from the same manufacturer, samples from the same lot would be assigned to the same class. A different data set would have to be gathered to answer these two questions. The data for the latter could be a subset of that needed to study the variation between manufacturers.

If sample history is not known and such an approach is not possible, a preliminary data analysis can be used to determine sample groupings. The results of sensory evaluations and/ or the chemical data can be used to define the classes. Principal components analysis or cluster analysis can be used to search for groupings in the samples. If groupings are found, it must be determined if they make sense in light of the purpose of the study.

B. DATA COLLECTION

Chemical data are not well behaved in the statistical sense. This situation is a result of the measurement techniques used for sample characterization, biochemical, process, or formulation induced variations, and the inability to properly control all experimental variables during data collection.

Modern analytical techniques (e.g., chromatography) are capable of generating a large number of variables characterizing each sample. This results in a low sample-to-variable ratio (too few degrees of freedom), preventing the application of parametric techniques. A systematic reduction in the number of variables characterizing each sample by selection of the most discriminating one is required to minimize this problem.

Complex chemical variations caused by plant physiology and the processing/formulation of flavor systems result in *collinearities* or *correlations* between variables. These variable-variable relationships must be identified as a part of the variable selection process. Most discriminant analysis techniques do not inherently yield information about the interrelationships between *variables*. They are designed to find the combinations of variables that discriminate between different classes of *samples*. Since highly correlated variables contain redundant information, they will yield similar results in a discriminant analysis. Thus, if a variable is a good predictor of class membership, all variables that are correlated to it will also be good predictors. To use two or more correlated variables to define the discriminant model does not yield additional useful information. The best combination of variables for discriminant analysis includes variables that represent independent measures of product similarities and differences.

Many times the data are not collected according to a strict experimental design. It is difficult to control variables across a series of samples that may be derived from different sources, that is, the data cannot always be gathered under conditions where the effect of a single variable is isolated from other effects.

These characteristics represent an undesirable situation for traditional statistical techniques and limit their application. The techniques commonly applied to analyze chemical data must be capable of operating under these conditions and extracting the signal from the noise. As a result, the most useful techniques are often those that are not based on statistical parameterization of the data.

Sensory data are similar to chemical data in these respects. A major difference between chemical and sensory data is that it is easier to collect enough sensory data points that the statistical requirements can be met. As a result, parametric techniques can often be applied to sensory data. For example, it would not be unreasonable to collect data for five samples from ten panelists in triplicate, that is, each panelist sees each sample three times. Such a data set would have $5 \times 10 \times 3$ degrees of freedom, and thus could be modeled by several variables without fear of invalidating any of the statistical assumptions.

It would be very time consuming and difficult to collect the corresponding chemical data for each of the 150 samples used for the above sensory evaluations. When the chemical and sensory data are to be merged, the sensory information must be pared down to the same number of samples as contained in the chemical data. This is usually done by averaging. The net result is a reduction in the number of degrees of freedom and thus limitations on the application of parametric techniques. Under these circumstances, nonparametric techniques must be applied to the sensory data as well.

As a rule of thumb, each sample class should have at least five members and the ratio of the number of samples to variables should be greater than or equal to three.[1] This is not the number of variables that are measured, but the number used simultaneously for calculation of the discriminant function or model. It is acceptable to begin a study with a minimum number of samples and to expand the data set as the study progresses. When working with a small data set, one must be aware of the limitations for the applications of the various techniques and the possibility for chance correlations. More samples help to better define the data structure and add confidence to the results obtained. With the above points in mind, one begins to see the need for a strategy to find the most discriminating variables.

C. FEATURE SELECTION AND MODEL DEVELOPMENT

The preprocessing of data is necessary to transform the data into meaningful quantities prior to the data analysis. The preprocessing of chemical data is described in detail in Chapter 4. Various techniques for preprocessing sensory data are covered in Chapters 6 and 7. Once the data have been preprocessed to calibrate the raw responses and to remove any artifacts due to differences in scale, the data analyses can be performed.

Many different schemes exist for *feature selection*. The main goals are to identify the most discriminating variables and to reduce the number of variables by eliminating redundancies. Practical consequences of feature selection include a potential reduction in the number of measurements that need to be made for sample characterization, an indirect evaluation of the experimental design, and an improved understanding of the basic chemical and sensory differences between samples.

Feature selection schemes can be univariate, multivariate, or a combination of the two. Univariate schemes look at each variable individually and apply a predefined criteria to make the decision about the discrimination ability of the variable. Multivariate schemes test a combination of variables for discrimination ability. It is often necessary to use a combination of the different approaches to ensure that all appropriate information has been found in the data set. The ultimate set of features for the discriminant analysis should contain variables which represent independent sources of variation.

Model development is inherently related to feature selection in that the selected variables are tested for their discrimination ability in order to identify the combinations that yield the best results. Feature selection and model development are accomplished by the use of a *training set*. A training set is a database that meets the criteria described above. It is used for the development of the discriminant model. The results of the analysis of the training set are expressed as *classification accuracy*. The percent correct classification reflects the ability of the selected features *and* the chosen discriminant analysis technique to separate the different classes within the training set.

1. Univariate Feature Selection

Calculation of correlation coefficients between all variables will identify the highly corre-lated ones and allow the user to select a representative from each group for further testing. This approach identifies redundant or collinear variables, but does not yield any indication of the discrimination ability of a variable. Correlation analysis must be combined with other univari-ate procedures for feature selection.

When enough samples are available, the *students t-test* can be used to gauge the discrimi-nation ability of each variable individually by testing for significant differences in the means of two classes. When fewer samples are available, nonparametric tests such as the Mann-Whitney U-test or the Wilcoxon T-test can be used.[2]

Another univariate approach is to calculate the ratio of the between-class variance to the within-class variance for each feature.[2] This is an F-test to determine if the ratio represents significant differences in the between- and within-class distributions.[3] Variables with large values for this ratio represent large differences in class means and small intraclass variance and are more likely to be good discriminators. A direct test for this condition is described by Massart et al., where the ratio of the difference in the class means to the pooled standard deviations for the two classes is calculated.[2]

To further enhance the performance of a variable, weighting schemes can be used.[2,4,5] Weights account for unequal variable variances. Variables with larger between-class vari-ances recieve larger weight in the calculations and thus have more influence on the results. Weighting schemes should be used only with larger data sets as the weighting tends to exaggerate the between-class differences.[1]

2. Multivariate Feature Selection

The schemes described above are univariate in nature, that is, they look at one variable at a time. The univariate approach to feature selection runs the risk of missing important information contained in variable combinations. Put another way, one variable may not be able to discriminate between several different classes, but the combination of that variable with others may yield the desired result.

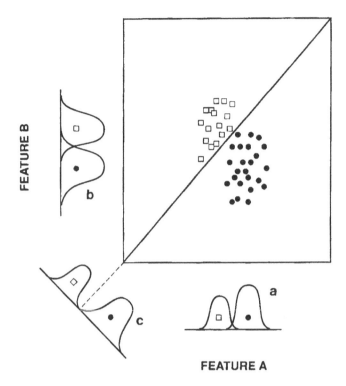

FIGURE 2. Hypothetical data space defined by variables A and B. The
frequency distributions of classes 1 and 2 on (a) variable A, (b) variable B,
and (c) the discriminant function. Overlap of the classes occurs for both
variables. Resolution is obtained by the discriminant function.

This concept is demonstrated graphically in the variable-variable *scatterplot* shown in
Figure 2. In this plot, variable A is plotted against variable B. The individual points represent
the different samples. Samples of the same class are represented by the same symbol. The
relationships between the different classes can be seen in this plot. Projection of the samples
onto the variable A axis shows that variable A does not completely resolve the two classes.
The same is found for variable B. There is overlap in the distributions for the two classes on
each variable. When the partial discrimination ability of the two variables is combined,
resolution of the classes is possible. The separation can be accomplished in this case by a linear
combination of the two variables, as demonstrated by the line or by a geometric analysis of
the distances (i.e., spatial relationships) between the class clusters. Both approaches are valid
and form the basis for different groups of discriminant analysis techniques. By analogy to this
two-dimensional example, such an analysis can be extended to more variables and more
classes to demonstrate the utility of using multiple variables in combination to address a
complex multiclass problem.

Principal components analysis, as described in Chapter 11 and Reference 6, can yield
useful information about variable-to-variable correlations, variable-to-sample relationships,
and sample-to-sample comparisons. As opposed to the stepwise procedures described above,
this approach allows one to check the correctness of class assignments, to identify variables
that are most likely to yield discrimination, and to select representative ones from each
correlated group of variables in one analysis. This approach is not based on the prior
classification of samples. It provides an unsupervised representation of the information
contained in the data. In effect, visual examination of the principal components loadings and

scores yields a discriminant analysis. After feature selection by this approach, the discriminant analysis technique would have to be used to determine the actual classification and prediction ability for a chosen set of variables. Schemes that use principal component data reduction simultaneously for feature selection and discriminant analysis have been developed. They will be discussed along with the other discriminant analysis techniques.

Other multivariate schemes are also useful for finding the best sets of discriminating variables. While somewhat computation intensive, many combinations of variables can be tested to look directly for multivariate discrimination ability. These stepwise procedures are generally implemented in forward or reverse search formats. A forward search systematically evaluates all possible combinations of variables by starting with lower dimensions and working to higher dimensions. For example, the best two-dimensional combination might be found initially. A third variable is then sought that improves the discrimination ability of the two-dimensional combination. The best three-dimensional combination can then be used as the basis for evaluating four variables simultaneously. The process continues until 100% classification ability is obtained for the training set. For reverse searches, the analysis begins with multidimensional combinations and systematically eliminates nonuseful variables. The forward search is usually more computationally efficient, since it allows one to find the minimum number of variables needed for discrimination. The major drawback for both is the large number of possible variable combinations that must be tested. On the other hand, these approaches yield the desired multivariate effects and ensure that potentially useful combinations of variables are tested.

D. VALIDATION OF RESULTS

The last step is to test the general validity of the discriminant model. Additional data are required for this test. Members of the *prediction set* or *test set* belong to the same classes as those in the training set, but represent new data that were not used in the development of the discriminant function. The variable combinations obtained from the analysis of the training set are used to predict the class memberships for the test set. Results for this stage of analysis are reported as *prediction ability*, expressed as the percent correct predictions for the test set.

When the number of samples available for training and prediction is limited, feature selection and validation can be combined into one step. There are procedures that can be used to estimate the prediction ability from a single data set. These approaches do not divide the data into training and prediction sets as defined above. Instead, the data set is temporarily segmented, part is used as the training set for the development of the discriminant model, while the rest is used as a test set. The segmentation procedure is repeated as many times as necessary so that each sample is used in combination with other samples in both the training and test sets. This approach is sometimes referred to as *cross validation*.

When the test set consists of only one sample at a time, it is called the *leave-one-out* approach. Each sample is removed, in turn, from the data base and its class membership is predicted based on the remaining samples. The results for the prediction on each sample are combined to yield the overall estimated prediction ability. The leave-one-out method provides the largest possible training set for comparison.

When several samples are simultaneously segmented for testing, it is called a *jackknife* procedure. The data set can be segmented randomly or in a more regular fashion. For example, the first n samples are used initially as the test set. After prediction, these samples are returned to the tranining set and the next n samples are removed for testing in the next iteration. A modification of this scheme is to remove every nth sample for the test set. The sequence is repeated until each subset within the whole data set has been tested. As for the leave-one-out procedure, the results of each iteration in the jackknife procedure are combined to yield an overall estimated prediction ability.

The use of a test set or one of the segmentation procedures to obtain an estimate of the prediction ability is necessary to ensure that the discriminant model developed is generally applicable to all related samples. Use of only the classification results from the training set may yield an overly optimistic estimate of the power of the discriminant function. A significant difference between the *classification accuracy* for the training set and the *prediction ability* for the test set suggests that the training set is not a representative sampling of the true sample space. More samples need to be included in the training set to accurately describe the sample data distributions.

III. DISCRIMINANT ANALYSIS TECHNIQUES

All discriminant analysis techniques are based on the assumption that members of the same class are more similar to each other than to members of other classes, and that the similarities and differences are reflected in at least some of the measured variables. Several different general methodologies exist for the comparison of samples in a multidimensional space. While different specific algorithms exist for each technique within a given methodological group, all the techniques within that group operate on essentially the same principles. Rather than discussing the many different techniques, the concepts for each methodological group will be described. Detailed discussions are available in numerous literature references.[1,2,5-9]

A. LINEAR DISCRIMINATION
In the example given in Figure 2, a line could be drawn between the *clusters* of points for each class. When two classes are linearly separable, techniques that are similar to linear regression can be used to define discriminant functions.

1. Linear Discriminant Analysis (LDA) and Stepwise Linear Discriminant Analysis (SLDA)
LDA and SLDA are parametric techniques. They assume multivariate normal distributions and thus can be applied to large data sets. The main drawbacks of these techniques are the assumption of linear separation of classes and the fact that a separate discriminant function is often required for each pair of classes. Stepwise analysis can be implemented in either forward or reverse search modes to find the optimum combination of variables (i.e., feature selection).

The general form of the discriminant function is given in Equation 1:

$$DS = a_1X_1 + a_2X_2 + a_3X_3 + ... + a_nX_n \tag{1}$$

The discriminant score, DS, is calculated for each sample and used to determine the class to which the sample belongs. The analysis consists of solving for the coefficients, a_i, that yield the highest classification accuracy for the training set. In a manner analogous to the procedure that least squares regression uses to minimize the sum of the squares of residuals, LDA maximizes the between-class to the within-class variance. A linear decision surface is determined that defines the boundary between classes. In two dimensions, the boundary is a line as in Figure 2. For three dimensions, the boundary is a *plane*. For more than three dimensions, the surface is a *hyperplane* of $(n - 1)$ dimensions where n is the number of variables used in the discriminant model.

If the variables, X_i, are standardized prior to analysis, DS is standardized and the *cut off* for class separation will be zero. Positive values for DS will imply membership in one class, while negative values will belong to the other class. Standardizing the variables prior to analysis also allows one to attach some significance to the discriminating power of the variable according to the absolute magnitude of the coefficient. Another measure of the discriminating power of a variable is the absolute correlation of the variable with DS.

When multiple classes are to be tested, each pairwise discriminant function is evaluated sequentially to determine classification. This sequential approach involves several decisions relative to the predefined cutoffs and can be quite complex. A more simplistic approach that works as well is to use the square of the Euclidean distance or Mahalanobis' D^2 procedure to measure the distance of each sample's DS to the mean for each class.[7]

LDA and SLDA have been widely used for the analysis of sensory data where it is relatively easy to gather large data sets. Application of these techniques to chemical data has been less successful due to the characteristics of chemical data mentioned earlier. In either case, the possibility of chance correlations, especially with smaller data sets must be considered.

2. Linear Learning Machine (LLM)

The linear learning machine is a nonparametric linear classification technique that belongs to a group of techniques called threshold logic unit (TLU) methods. As implied by the name, classification decisions are based on comparison of the calculated result to a predefined threshold value. If the result exceeds the threshold, the sample belongs to one class. Otherwise, the sample is put into the other class. The threshold for the LLM is usually zero. Thus, the decision is similar to that for LDA with standardized variables. Like LDA, a linear equation is developed in order to classify samples. Unlike LDA, the equation is not defined by a statistically derived criterion.

The general equation for LLM is of the form:

$$s = W \cdot X = |w|\,|x|\cos\Theta \tag{2}$$

where the dot product between the weight vector, W, and the data vector, X, is calculated. Θ is the angle between the data vector and the weight vector. The purpose of the analysis is to determine the weight vector, W, that provides for correct classification. This effectively creates a decision surface of dimension $n-1$, where n is the number of variables in the data vector. W is defined as perpendicular to the decision surface. The decision surfaces are the same as for LDA. When the weight and data vectors lie on the same side of a decision surface, $\Theta < 90°$, $\cos \Theta > 0$, and the result is positive. When they are on opposite sides, $\Theta > 90°$, $\cos \Theta < 0$, and the result is negative. This polarity of the calculated scalar provides the basis for classification.

In practice, the dot product is calculated as the sum of the products of the elements of the two vectors:

$$s = w_1x_1 + w_2x_2 + w_3x_3 + ... + w_nx_n + w_{n+1}z \tag{3}$$

where z is an additional dimension added to all samples and usually given the value of unity. This positions the decision surface through the origin.

An analysis is performed by determination of the weight vector, W, yielding the scalar, s, that correctly classifies the samples in the training set. Usually an arbitrary weight vector is selected initially; it is adjusted iteratively. Each time an incorrect classification is obtained, the decision surface is reflected (by the same amount that it was in error) through the misclassified point to produce a weight vector that correctly classifies that sample. As this correction process continues, the algorithm learns where the distinction between the two classes exists. Multiclass problems are handled as a series of binary decisions with a different decision surface (discriminant function) used for each pair of classes.

LLM is the nonparametric equivalent of LDA. The LLM has been used more extensively in chemistry. Several different approaches exist for implementation of the procedure.[5,8,9] Like LDA, the assumption of linear separation poses limitations. Since the decision surface is determined through an error correction feedback procedure, the result is often dependent on

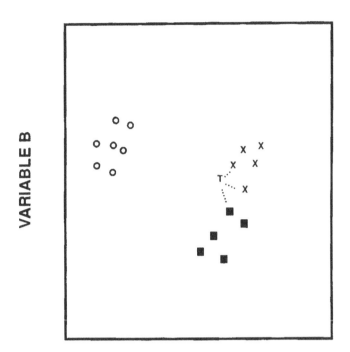

VARIABLE A

FIGURE 3. K-nearest neighbor analysis. For both training and prediction, the distances between the test sample T and all other samples are compared to determine the K-nearest neighbors. For K = 3, shown by the dashed lines, the test sample would be assigned to category X based on the majority vote of the three nearest neighbors.

the order in which samples were presented. A nonunique solution can be obtained. Like the other discriminant analysis techniques, the outcome is a function of the samples used in the training set. As with LDA the probability of chance correlations must be considered.[10,11]

B. K-NEAREST NEIGHBOR ANALYSIS (KNN)

The nearest neighbor discriminant analysis scheme mathematically implements the geometric analysis of a multidimensional feature space. In such an approach, hypothetical coordinate systems are established as exemplified in feature-feature scatterplots. Each feature defines a dimension in the feature space. There is no limit to the number of dimensions that can be handled mathematically. The limiting factor is the sample-to-feature ratio criterion mentioned earlier. The measure of similarity between samples is the *distance* between them in the multidimensional space. The Euclidean distance is usually used as the metric of similarity:

$$D = [(X_{ia} - X_{ja})^2 + (X_{ib} - X_{jb})^2]^{1/2} \qquad (4)$$

The calculation defined in Equation 4 is the distance between samples i and j based on features a and b as in Figure 3. For more than two dimensions, an additional term for each feature is used in the calculation.

This approach is conceptually easy to grasp since it is analogous to visual inspection of scatterplots. Classification is based on the proximity of samples. The technique is referred to as K-Nearest Neighbor (KNN), where K represents the number of nearest neighbors that are polled to determine the class membership of the test sample. The test sample T in Figure 3 would be classified as an x based on the majority vote of its three nearest neighbors as indicated by the dotted lines.

K is usually equal to 1 or 3; K = 1 yields a minimum of information about class structure, whereas larger values for K give an implicit evaluation of the density of the class cluster and the spatial closeness of different classes. For example, high classification accuracy for K = 3 or 5 suggests that the cluster for a class is tight (i.e., the class is homogeneous) and that the distance between classes is generally larger than the spread within the class, that is, there is a large ratio of the between class variance to the within-class variances. Good classification results for K = 1 with poor results when K = 3 or 5 suggests that the ratio of the between- to within-class variances is small. This may be caused by the presence of subsets of samples (i.e., the class is nonhomogeneous) in the class where performance has decreased. The reevaluation and possible redefinition of class membership is warranted in such a situation. For large values of K, good results are obtained only for tightly clustered, well separated classes. For classes that are spatially close to each other, too large a value of K can prevent discrimination of samples near the boundary region between classes. The optimum value for K can usually be identified with a few iterations of mathematical analysis and visual inspection of feature space plots.

KNN analysis makes no assumptions about the structure of the data such as linear separation. It can be applied to several classes simultaneously and can be used with smaller sample-to-variable ratios than the linear discrimination techniques. The solution obtained is unique and is not dependent on the order of sample comparison. It is not as susceptible to outliers as the linear discrimination techniques.

The KNN technique has a bounded classification risk. Most of the classification information is found in the first nearest neighbor, and any other classification technique can at most double the performance of KNN.[5,12] The KNN technique is susceptible to differences in data scales and requires the application of scaling techniques to remove these differences. Autoscaling or standardization is most commonly used to circumvent this problem. Another drawback to the KNN approach is that it can be computationally intense for a relatively large data set, as the distances between the test sample and all others must be calculated and compared to determine the nearest neighbors.

Forward searching feature selection, using the leave-one-out procedure, is a very powerful discriminant analysis tool when combined with the KNN technique.[13,14] For example, the discrimination power of each variable can be tested individually. The best features (say, ten) from the one-dimensional analysis can be tested in higher-dimensional combinations to find the best features for multiple class discrimination. The features from the best two-dimensional combinations can then be tested in three dimensions and so on. The beauty of this approach is that all possible combinations of features can be evaluated. Since only the best features are retained from the previous analysis, fewer features are used in each subsequent analysis, thereby minimizing the number of feature combinations that must be tested. This can become significant as the number of possible combinations of features to be tested is potentially very large.

C. CONTINUOUS MODELS

Multivariate discriminant analysis techniques are similar to statistical procedures in that they provide a measure of the confidence in the similarities and differences between samples. For the techniques discussed so far, the measure of confidence is the prediction ability. Higher prediction abilities imply greater confidence in the identified sample differences. Nonsupervised, continuous modeling techniques, such as principal components analysis, (PCA), do not generally yield measures of confidence in the observed similarities and differences. Nevertheless, we do use the results of a PCA (i.e., the sample scores) to compare samples. Visual

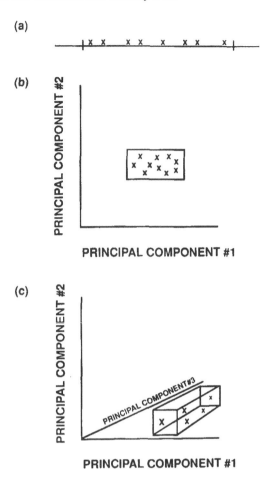

FIGURE 4. SIMCA class models: (a) one-dimensional, (b) two-dimensional, and (c) three-dimensional class boundaries.

inspection of PCA score plots has been mentioned earlier as a means of checking class membership. If the sample scores are used as the input to a discriminant analysis procedure, we have combined the distinctly different methodologies. Ideally, such a combination takes advantage of the desirable attributes of each technique and provides an advantage over the use of each procedure separately.

1. SIMCA

It is possible to develop an estimate of confidence when a discriminant approach is combined with principal components analysis. The SIMCA (*soft independent modeling of class analogy*) technique developed by Wold is just such a technique.[15] Each class is separately modeled by PCA. This is called a disjoint analysis. The number of principal components (npc) needed to model each class is usually determined by *cross validation*. Confidence limits based on class variances and centered around the mean of each class are established. The regions specified by the confidence ranges define the npc-dimensional space occupied by that class. When npc = 1, the class space is defined by the range of values on a line. For npc = 2, the region centered around the class mean in the two-dimensional plane defines the class space. A "box" defines the class space for npc = 3. These ideas are represented graphically in Figure 4. Unknowns are predicted by projection into the data space and determination of the overlap with the defined class spaces.

The SIMCA technique extends the use of the standard PCA technique to the discriminant analysis category by analyzing the variances associated with the PCA model. As discussed in Chapter 11, the PCA model is composed of two types of variances. The first type of variance represents the *signals* in the data. It describes the interrelationships between the variables and allows the comparison of samples for similarities and differences. This is the basis for the SIMCA class models. The second type of variance is *noise* or error left over after definition of the model. This is also referred to as the *residual* variance. Analysis of the residual variance provides the statistical tests for "goodness of fit", as well as the discrimination decision for the SIMCA method. Both the samples and variables are characterized by the analysis of residual variances.

Comparison of the residual variances for all relavent features across all samples in a class to that for an individual sample assigned to that class allows identification of outliers (i.e., incorrectly assigned class membership) by means of an F test. If the residual variance for the class is similar to that for the test sample, the test sample is not significantly different from the other members of that class. If the test sample residual variance is significantly greater, it is considered to be an outlier. The same approach is used to make predictions on an unknown. An unknown is assigned to the class whose model best describes it, that is, the unknown is not considered to be an outlier for that class.

By looking at the residual variance for variables, the *modeling power* of a variable can be calculated as a function of the ratio of the error variance to the modeled variance for each variable individually. As the ratio approaches 0.0, the variable contributes significantly to the model; as it approaches 1.0, the variable is of little consequence to the model. The *discriminating power* of a variable can be determined from a more complex expression involving residual variances.

The SIMCA technique combines both discriminant analysis and continuous modeling methodology. Correlations among variables (i. e., colinearities) are used to advantage since the effect of averaging the information in related variables makes the model more robust. Measures of confidence beyond the normal discriminant analysis prediction abilities are also obtained. For these reasons, SIMCA is considered to be the method of choice for many applications.

IV. SUMMARY

Discriminant analysis techniques offer the simplest level of sample comparison. By identifying those variables that represent differences between classes, one can begin to develop a chemical rationale for sensory changes in products. Mathematically significant differences do not necessarily imply *cause and effect* relationships. Further testing is required to establish cause and effect relationships between the observed sample differences in the chemical and sensory data. If true cause and effect relationships cannot be established, reproducible differences in composition that correlate with sensory changes can be used as *indicator* variables. Such variables are indicative of specific sample characteristics and can be used to predict certain behaviors even though they are not directly the cause for sensory differences. One must be aware of the potential risks involved when using indicator variables.

Many of the parametric techniques developed in other disciplines are not applicable to chemical and sensory data. The techniques described above represent those most often applied in the study of chemical and sensory data. Many other techniques exist and are being developed. As newer techniques are tested in more applications, they may gain popularity and eventually prove to be the techniques of choice.[16]

V. APPLICATION OF DISCRIMINANT ANALYSIS TO THE TEA DATABASE

Discriminant analysis was applied to the chemical profiles collected in Chapter 8. Similar analyses could be performed on the sensory data as well. The purpose of the discriminant analysis was to identify those chemical compounds that could be used to discriminate between the tea samples. The problem was set up using seven different classes of samples. At least five replicate analyses for each class were obtained. Several different feature selection stategies were used to search for the most useful features. The details of each step of the analysis sequence are described below.

The criteria for definition of classes was based on the tea manufacturer (product identification) primarily and on the method of preparation (number of tea bags, steeping time) for samples M, L, and T. These samples were all from the same manufacturer, but prepared differently to demonstrate concentration effects. The details of the sample preparation for all the tea samples are described in Chapter 8. This set of samples will then allow a comparison of different manufacturers' products that all fall into the category of orange-spice black teas and to look for the effects of variation in preparation. Differences in this data set could be caused by varieties of tea used, processing of the tea, and formulation with the orange and spice flavors.

Table 1 contains the averages and relative standard deviations for the replicate chemical analyses for each variable that was chosen from the chromatographic profiles. For the sake of simplicity, only those peaks in the chromatographic profiles whose identities were known are included in the data set. There were many more unknown peaks that could have been used. The standard deviations represent the variations caused by the sample preparation and the analytical technique. The reproducibility of sample preparation is determined by the variations in the steeping process and tea bag-to-tea bag differences. The data in Table 1 form the basis for a univariate t-test to determine variables that provide statistically confident discrimination. The t-tests must be done pairwise, that is, each pair of classes (21 possible combinations) must be compared on each variable. The results of t-tests are given in Table 2. It shows the variables for each pairwise comparison that discriminated at a confidence level of 95%. The last column shows the frequency of discrimination for each feature. Because of the relatively large differences in sample composition, it is possible to discriminate all the different pairs of samples. No single variable could discriminate all samples simultaneously. Only peak 20 came close to yielding complete discrimination as it was successful in 20 of the 21 possible comparisons. From this information, one might choose those peaks for further analyses that are above a certain frequency. For example, those peaks with frequencies of 16 or greater would yield discrimination in 16 of the 21 (about 75%) pairwise combinations. From this analysis, one would be interested in peaks (in order) 20, 31, 8, 23, 2, 5, 25, 18, 33, 6, 7, 17, and 22.

Table 3 shows the variable to variable correlations identified by PCA. This analysis also indicated that the assigned classes were capable of being discriminated by the chemical data. No outliers were detected.

From the information in Tables 1, 2, and 3, we can begin to understand the relationships between the samples but the sheer volume of information overwhelms us quickly. A simpler means of comparing samples and displaying interrelationships is desirable.

In addition to the univariate tests just described, KNN was used for multivariate feature selection and discriminant analysis. The ability of features to *simultaneously* discriminate all seven classes was tested. The data were autoscaled prior to the analysis. The algorithm used has the capability of forward-searching feature selection using the leave-one-out procedure. The sequence started with a one-dimensional (univariate) KNN analysis and retained the ten best features from this analysis for higher dimensional analyses. Two- and three-dimensional analyses were then performed to search for discriminating combinations of features. Tables 4 and 5 summarize the results of this sequence of analyses. The results reported in Table 4 were based on one-nearest neighbor (K = 1) and those in Table 5 for three-nearest neighbors (K = 3).

The one-dimensional analysis selected many of the same peaks as the t-test. The two-dimensional analysis showed that several combinations of peaks yielded 100% classification for K = 1. For K = 3, no two-dimensional combinations gave 100% accuracy. Three-

TABLE 1
Averages and Relative Standard Deviations for Each Class

Feature	Name	B AVG	RSD	R AVG	RSD	Q AVG	RSD	M AVG	RSD	L AVG	RSD	T AVG	RSD	S AVG	RSD
1	Gallic acid	21.3	24	12.1	5	27.9	24	8.68	14	22.1	11	19.7	18	17.9	15
2	Theobromine	11.9	13	4.67	6	15.3	13	4.74	4	11.1	4	9.05	4	10.6	4
3	Theophylline	0.71	50	2.06	8	1.73	17	0.35	13	0.47	65	0.53	34	0.49	21
4	EGC[a]	12.7	9	8.40	36	15.2	30	4.79	26	15.5	32	9.51	39	8.78	35
5	Caffeine	161.	6	99.1	3	146.	17	56.9	7	136.	4	113.	3	129.	4
6	Epicatechin	24.7	14	10.9	24	13.1	36	1.89	36	9.45	25	6.36	10	7.08	29
7	EGCG[b]	156.	21	55.9	16	120.	16	14.3	24	95.2	25	72.9	17	70.1	16
8	ECG[c]	147.	15	49.5	18	113.	11	25.7	24	77.6	23	50.3	21	51.5	14
9	Citric acid	58.4	16	72.3	8	57.8	18	16.3	8	49.3	24	37.7	18	37.3	13
10	Malic acid	11.0	19	19.0	22	9.14	27	9.03	93	10.7	27	11.6	26	9.14	73
11	Succinic acid	474.	17	381.	6	709.	6	202.	8	433.	12	389.	14	397.	5
12	fumaric acid	0.46	10	0.47	9	0.54	20	0.21	26	0.37	14	0.40	19	0.30	12
13	Ethyl butyrate	0.00	—	0.35	11	0.00	—	0.00	—	0.00	—	0.00	—	0.01	20
14	α-Pinene	0.03	53	0.06	23	0.00	—	0.00	—	0.00	—	0.00	—	.03	10
15	β-Myrcene	0.11	52	0.18	20	0.00	—	0.01	99	0.01	13	0.02	59	0.17	7
16	Ethyl hexanoate	0.00	—	.07	27	0.00	—	0.00	—	0.00	—	0.00	—	0.00	—
17	Octanal	0.01	39	0.05	27	0.00	—	0.00	—	.01	60	0.01	80	0.02	31
18	Limonene	8.84	52	10.9	21	0.05	17	0.67	67	1.10	9	1.36	52	19.7	8
19	γ-Terpinene	0.25	17	0.26	31	0.12	17	0.35	28	0.24	13	0.22	20	0.22	28
20	Linalool	0.47	17	1.67	9	0.23	17	0.51	11	0.69	9	0.59	6	0.96	4
21	Cineole	0.13	34	0.11	51	0.10	25	0.10	25	0.12	39	0.12	60	0.11	69
22	4-Terpineol	0.06	7	0.10	14	0.04	19	0.06	8	0.09	7	0.08	9	0.06	17
23	α-Terpineol	0.13	33	0.54	7	0.03	55	0.07	11	0.08	34	0.08	21	0.80	5
24	Neral	0.04	39	0.18	13	0.00	—	0.00	—	0.00	—	0.00	—	.16	7
25	D-Carvone	0.03	39	0.10	13	0.02	21	0.04	8	0.06	14	0.05	16	0.08	8
26	Geranial	0.06	24	0.23	25	0.02	90	0.09	80	0.07	67	0.06	53	0.17	11
27	Cinnamic aldehyde	0.26	24	0.17	25	0.08	68	0.17	19	0.16	15	0.15	22	0.15	20
28	Perialdehyde	0.07	17	0.10	22	0.04	76	0.09	16	0.09	11	0.08	17	0.13	12
29	Undecyclic aldehyde	0.05	58	0.08	14	0.04	75	0.09	19	0.07	13	0.07	21	0.09	16
30	Neryl acetate	0.00	—	0.01	24	0.00	—	0.00	—	0.00	—	0.00	—	.05	14
31	Eugenol	12.5	15	8.95	4	2.08	11	9.60	16	23.0	7	18.6	3	18.6	2
32	Dodecanal	0.12	6	0.04	25	0.00	—	0.00	—	0.00	—	0.00	—	.36	19
33	β-Caryophyllene	1.07	31	0.07	8	0.02	78	0.05	32	0.06	17	0.07	21	2.62	6
34	Valencene	1.19	71	0.36	10	0.00	—	0.00	—	0.00	—	0.00	—	1.73	99

[a] EGC = epigallocatechin.
[b] EGCG = epigallocatechin gallate.
[c] ECG = epicatechin gallate.

dimensional analyses yielded the desired classification performance for K = 3. The difference in results for K = 1 and K = 3 indicates that some of the classes are very close to each other or that subsets exist within those classes.

The multidimensional analysis results demonstrate a point that was made earlier. Univariate feature selection may not find which features are best in combination with others. In this case, the univariate tests (t-test and one-dimensional KNN) indicated that peak 20 would be the best single feature for discrimination. When used in combinations, peaks 5 and 31 were more powerful. This is demonstrated in Figures 5 and 6. The spatial closeness of samples B, L, M, and T on feature 20 and the similarity of L, T, and S on feature 31 is apparent. In addition, there appears to be a possible subset of sample Q on feature 5. These characteristics help to explain the differences obtained when K = 1 and K = 3. The value of the second dimension, to improve the discrimination, can be readily seen in these figures. Figure 7 is a three-dimensional plot of features 5, 20, and 31. The additional resolution of the classes, allowing K = 3 to be used for discrimination, is visible.

TABLE 2.
Pairwise t-Test Discrimination Matrix

Feature	B-R	B-Q	B-M	B-L	B-T	B-S	R-Q	R-M	R-L	R-T	R-S	Q-M	Q-L	Q-T	Q-S	M-L	M-T	M-S	L-T	L-S	T-S	Frequency
1	+		+				+	+	+	+	+	+		+	+	+	+	+		+		14
2	+	+	+		+		+	+	+	+	+	+	+	+	+	+	+	+	+		+	17
3	+	+	+				+	+	+	+	+	+		+	+	+		+				13
4	+		+	+	+	+	+	+	+			+		+	+		+	+	+	+	+	14
5	+	+	+	+	+	+	+	+	+	+	+	+		+	+		+	+	+			17
6	+		+	+	+	+		+	+	+	+	+		+	+		+	+				16
7	+	+	+	+	+	+	+	+	+	+	+	+		+	+		+	+				16
8	+		+	+	+	+	+	+	+	+	+	+	+	+	+		+	+		+	+	18
9	+	+	+	+	+	+	+			+	+	+			+			+				15
10	+						+	+														6
11	+	+	+	+	+		+	+			+	+	+	+	+	+	+			+	+	13
12						+		+	+										+	+	+	15
13						+	+	+	+									+		+	+	11
14	+		+	+	+		+	+	+	+	+		+	+	+			+		+	+	14
15	+	+	+	+	+			+			+	+		+	+	+		+		+	+	15
16	+					+	+	+	+	+	+	+	+		+							6
17	+	+	+	+	+	+	+	+		+	+	+	+	+	+	+	+	+		+	+	16
18		+				+	+	+		+		+	+	+	+	+	+	+		+	+	17
19															+							10
20	+			+	+	+	+	+	+		+		+	+	+	+			+	+	+	20
21																						0
22	+	+	+	+	+	+	+	+	+	+	+	+	+	+	+	+	+	+	+	+	+	16
23	+	+	+	+	+	+	+	+	+	+	+	+	+	+	+	+	+	+	+	+	+	18
24	+		+	+	+	+	+	+	+	+	+	+	+	+	+					+	+	14
25	+	+	+		+	+	+	+	+	+	+	+	+	+	+	+		+	+	+	+	18
26	+	+		+		+	+	+			+	+	+	+	+					+		11
27	+	+	+	+	+	+	+					+	+	+	+			+				11
28	+	+	+	+		+	+	+				+	+	+	+	+	+	+				14
29			+	+		+	+				+	+	+	+	+							9
30	+	+			+	+	+	+		+		+	+	+	+	+	+	+	+	+	+	11
31	+	+	+	+	+	+	+		+	+	+	+	+	+	+			+		+	+	19
32	+	+	+	+	+	+	+	+	+	+	+	+	+		+			+		+	+	15
33	+	+	+	+	+	+	+	+	+	+	+	+		+	+			+		+	+	17
34	+	+	+	+	+	+	+	+	+	+		+	+	+	+					+	+	10

Note: A + indicates discrimination at 95% confidence level.

TABLE 3
Groups of Correlated Features Defined by
PCA

Group no.	Features
1	1, 2, 4, 5, 6, 7, 8, 11, 12
2	3, 9
3	10, 27, 30, 32, 33, 34
4	20, 25, 26
5	13, 14, 15, 16, 17, 18, 23, 24
6	21
7	19, 22, 28, 29, 31

TABLE 4
KNN (K = 1) Results for the Ten Best Features and Combinations

Features One dimension	%Correct classifications							
	B	R	Q	M	L	T	S	Overall
20	57	100	100	20	80	67	100	74
18	57	50	100	40	100	67	100	72
31	71	83	100	40	100	33	40	67
23	86	100	80	40	20	33	100	67
5	71	100	20	100	20	100	40	67
15	57	33	80	20	80	100	60	60
33	100	33	80	40	40	0	100	56
6	100	33	20	100	0	67	60	56
1	29	100	20	100	40	67	40	56
3	43	83	80	60	40	33	40	54
Two dimensions								
5, 18	100	100	100	100	100	100	100	100
5, 31	100	100	100	100	100	100	100	100
5, 33	100	100	100	100	100	100	100	100
18, 31	86	100	100	100	100	100	100	97
23, 31	100	100	100	80	100	100	100	97
5, 15	86	100	100	100	100	100	100	97
15, 31	86	100	100	100	100	100	100	97
31, 33	100	100	100	80	100	100	100	97
18, 20	100	100	100	80	100	83	100	95
5, 20	100	100	100	100	80	100	80	95

What is the potential impact of these chemical differences on sensory characteristics? The discussion will center around the top 15 discriminating compounds and be broken down into the 3 main flavor vectors in these samples, i.e., the tea flavor itself, the orange, and the spice flavors. The simultaneous comparison of all samples will be considered. A detailed comparison of each pair of samples could also be performed.

The methyl xanthine compounds, measured by the HPLC profile, would account primarily for variations in astringency and bitterness. The variations in the epicatechin (compound 6), EGCG (compound 7), and ECG (compound 8) are related to each other and are primarily a reflection of the degree of fermentation of the tea leaves during processing[17] (for all samples prepared to the same brew solids level) or the concentration of the tea solids in the samples as prepared. The solids content of the tea solutions were monitored by refractive index as a part of sample preparation. They were found to be the same for the multiple preparations of each sample and relatively constant for all the samples, except M and T that were prepared differently to demonstrate concentration effects. Therefore, variations in these profiles would mostly be indicative of differences in tea processing except for samples M and T. Compounds

TABLE 5
KNN (K = 3) Results for the Ten Best Features and Combinations

Features One dimension	B	R	Q	M	L	T	S	Overall
20	71	100	100	60	80	67	100	82
5	86	100	20	100	40	100	40	72
23	100	100	80	80	0	17	100	69
18	57	67	100	20	80	50	100	67
33	100	83	80	0	20	0	100	56
31	100	83	100	0	100	0	0	56
6	86	67	0	100	20	0	0	54
3	71	100	60	60	20	0	0	46
15	71	0	80	0	80	67	0	44
1	43	100	0	100	0	0	0	36
Two dimensions								
5, 31	100	100	80	100	100	100	100	97
18, 31	86	100	100	100	100	100	100	97
5, 20	100	100	60	100	100	100	100	95
15, 31	71	100	100	100	100	100	100	95
20, 31	86	100	100	60	100	100	100	92
23, 31	86	100	100	60	100	100	100	92
5, 23	86	100	60	100	80	100	100	90
18, 20	86	100	100	80	100	67	100	90
5, 18	86	100	60	100	80	100	100	90
20, 33	100	100	100	80	80	67	100	90
Three dimensions								
5, 20, 31	100	100	100	100	100	100	100	100
18, 23, 31	100	100	100	100	100	100	100	100
20, 31, 33	100	100	100	100	100	100	100	100
5, 31, 33	100	100	100	100	100	100	100	100
23, 31, 33	100	100	100	100	100	100	100	100
3, 31, 33	100	100	100	100	100	100	100	100
15, 23, 31	100	100	100	100	100	100	100	100
5, 23, 31	100	100	80	100	100	100	100	97
18, 20, 31	86	100	100	100	100	100	100	97
5, 18, 31	86	100	100	100	100	100	100	97

6, 7, and 8 reflect the astringency characteristics of the teas. Caffeine (compound 5), which is derived by different mechanisms, would again reflect processing variations for equal brew solids as well as concentration differences in these samples. Caffeine is generally associated with bitterness. The levels of the methyl xanthines and caffiene also impact the *overall strength* of the samples.

Linalool (feature 20) is considered to be an indicator of tea quality. Higher levels are found in better teas.[18] Linalool was found to be one of the best discriminators, suggesting a wide range in the quality of the teas used in the various products. It is also an acid-catalyzed hydrolysis product of orange oil terpenes. Since most of the orange flavoring in these teas probably came from orange oils, linalool could also be introduced into the samples with the orange oils. The relative contribution of the two sources is not clear.

The major differences in the orange character of the samples are caused by the level and variety of orange oil added to the product. The variation in the limonene (compound 18) level, the major constituent of orange peel oil, does have some significance, but the variation in other compounds such as α-pinene (compound 14) and β-myrcene (compound 15) are minor. However, the hydrolysis products of these compounds, linalool (compound 20), α-terpineol (compound 23), D-carvone (compound 25), and 4-terpineol (compound 22) all tend to be good discriminators. This may be indicative of a large variation in the quality of the orange oils used

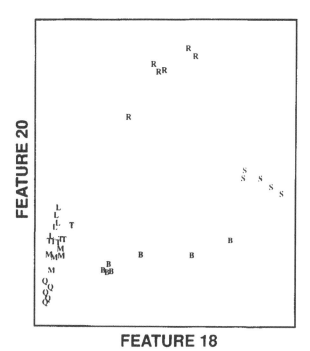

FIGURE 5. Two-dimensional scatter plot for features 18 and 20. Samples B, Q, R, and S are easily resolved while samples L, M, and T are partially overlapped lowering classification accuracy for K = 3.

in the various products. The flavor impact of these compounds is generally considered to be negative (from an orange flavor perspective) if the levels are too high. As mentioned previously, linalool is indigenous to tea so the cause for the variation in its level is not clear. Other compounds, derived from the orange oil are octanal (compound 17) and β-caryophellene (compound 33). Octanal is orange-like at low levels, but becomes fatty/soapy at too high a level. β-Caryophellene comes from the sesquiterpene fraction, as does valencene (compound 34), indicating variations in the heavier sesquiterpene fraction of the orange flavor composition. The sesquiterpene fraction is generally considered to be positive at moderate levels but spicy/peppery at too high a level. It also can change the mouthfeel of the product.

An important point to be considered is that the discriminant analysis, as performed, was searching for the best features for simultaneous discrimination of all seven classes. The mathematical criteria that were being met may not have reflected all the important characteristics in the data set. Compounds like ethyl butyrate (compound 13) and valencene (compound 34) can have an impact on the orange flavor character, but were not shown as good discriminators for all classes since they were only present in a few of the samples. The scientist must take into account the results of the discriminant analysis as well as other information (sensory thresholds) present in the data when drawing conclusions.

For the spice character, eugenol (compound 31) provided the best discrimination. Eugenol is probably responsible for the clove note in these samples. Most of the samples also contained cinnamic aldehyde (compound 27), but the variation in level was small. Therefore, most of the differences in spice impression were probably derived from the variation in the eugenol.

By looking at these differences and ranking the samples based on the levels of the key compounds mentioned above, one may be able to predict relative intensities of the samples on some of the different attributes. For the tea character, the order based on features 1 to 8 would be B > Q > L > S > T > R > M. The ordering obtained from the sensory analysis for tea taste is Q > L > B > T > S > M > R. From the analysis of the orange flavor compounds, the order

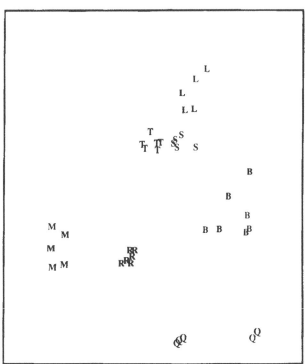

FEATURE 5

FIGURE 6. Two-dimensional scatter plot for features 5 and 31. All classes are more clearly resolved yielding better classification accuracy for K = 3.

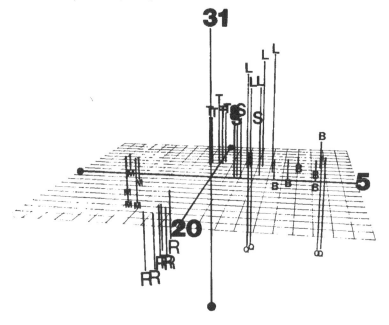

FIGURE 7. Three-dimensional plot for features 5, 20, and 31. Resolution of the different classes is even better when the third dimension is included.

would be R > S > B > T > L > M > Q. The orange aroma sensory description ranked the samples S > R > B > L > T > M > Q. For the clove taste, the sensory order was L > T > B > S > M > R > Q, compared to L > T > S > B > M > R > Q as determined by the eugenol level. If caffeine is considered to be the primary cause of bitterness in these tea samples, an order of B > Q > L > S > T > R > M is compared to the sensory ratings order of L > Q > B > S > T > R > M. These relative rankings are approximate and do not necessarily take into account the interaction of the different sensations. However, there is reasonably good agreement between the chemical and sensory data. Such similarities suggest the possiblity of cause and effect relationships. At this point, cause and effect has not been established. Additional experiments would be necessary to demonstrate the relationships.

The relative order of the samples on the above-mentioned flavor vectors are significantly different. Therefore, the products should be sensorially different and represent different combinations of the flavor sensations. These interactions could be synergistic or opposing (masking). The continuous modeling techniques discussed in Chapter 11 can take such information into account and begin to model such behavior.

REFERENCES

1. **Wold, S., Albano, C., Dunn, W.J., III, Edlund, U., Esbensen, K., Geladi, P., Hellberg, S., Johansson, E., Lindberg, W., and Sjorstrom, M.,** Multivariate data analysis in chemistry, in *Chemometrics — Mathematics and Statistics in Chemistry*, Kowalski, B.R., Ed., D. Reidel, Publishing Company, Dordrecht, The Netherlands, 1984.
2. **Massart, D.L., Vandeginst, B.G.M., Deming, S.N., Michotte, Y., and Kaufman, L.,** *Chemometrics: A Textbook*, Elsevier, Amsterdam, 1988, chap. 23.
3. **Caulcutt, R. and Boddy, R.,** *Statistics for Analytical Chemists*, Chapman and Hall, London, 1983.
4. **Forbes, R. A., Tews, E. C., Freiser, B. S., Wise, M. B., and Perone, S. P.,** Development of a novel weighting scheme for the K-nearest neighbor algorithm, *J. Chem. Inf. Comput. Sci.*, 26, 93, 1986.
5. **Sharaf, M. A., Illman, D. L., and Kowalski, B. R.,** *Chemometrics*, John Wiley & Sons, New York, 1986, chap. 6.
6. **Vogt, N. B.,** Principal component variable discriminant plots: a novel approach for interpretation and analysis of multi-class data, *J. Chemometrics*, 2, 81, 1988.
7. **Powers, J. J. and Ware, G. O.,** Discriminant analysis, in *Statistical Procedures in Food Research*, Piggot, J.R., Ed., Elsevier, London, 1986.
8. **Lachenbruch, P. A.,** *Discriminant Analysis*, Hafner Press, New York, 1975.
9. **Nilsson, N. J.,** *Learning Machines*, McGraw-Hill, New York, 1965.
10. **Lavine, B. K., Jurs P. C., and Henry, D. R.,** Chance classifications by non-parametric linear discriminant functions, *J. Chemometrics*, 2, 1, 1988.
11. **Lavine, B. K. and Henry, D. R.,** Monte Carlo studies of non-parametric linear discriminant functions, *J. Chemometrics*, 2, 85, 1988.
12. **Kowalski, B. R. and Bender, C. F.,** The k-nearest neighbor classification rule applied to nuclear magnetic resonance spectral interpretation, *Anal. Chem.*, 44, 1405, 1972.
13. **Carpenter, R. S., Burgard, D. R., Patton, D. R. and Zwerdling, S. S.,** Application of multivariate analysis to capillary GC profiles: comparison of the volatile fraction in processed orange juices, in *Instrumental Analysis of Foods, Vol. 2*, Charalambous, G., Ed., Academic Press, New York, 1983.
14. **Burgard, D. R. and Anast, J. M.,** Characterization of flavor systems—automation and multivariate analysis techniques, paper presented at ACS National Meeting, Symp. on Chemometrics, Anaheim, CA, September, 1986.
15. **Wold, S. and Sjostrom, M.,** SIMCA: a method for analyzing chemical data in terms of similarity and analogy, in *Chemometrics: Theory and Application*, Kowalski, B. R., Ed., ACS Symp. Ser. 52, American Chemical Society, Washington, D.C., 1977.
16. **Frank, I. E. and Friedman, J. H.,** Classification: oldtimers and newcomers, *J. Chemometrics*, 3, 463, 1989.
17. **Graham, H. N.,** Tea: the plant and its manufacture; chemistry and consumption of the beverage, in *The Methylxanthine Beverages and Foods: Chemistry, Consumption and Health Effects*, Spiller, G.A., Ed., Alan R. Liss, New York, 1984.

18. **Godwin, D. R.,** Relationships Between Sensory Response and Chemical Composition of Tea, Ph.D. thesis, University of Georgia, Athens, GA, 1984.

GENERAL REFERENCES

1. **Green, P.E.,** *Analyzing Multivariate Data,* Dreyden Press, Hinsdale, IL, 1978.
2. **Chatfield, C. and Collins, A.,** *Introduction to Multivariate Data Analysis,* Chapman and Hall, London, 1980.
3. **Martens, M., Dalen, G. A., and Russwurm, H., Jr., Eds.,** *Flavor Science and Technology,* John Wiley & Sons, Chichester, England, 1987.
4. **Martens, H. and Russwurm, H., Jr., Eds.,** *Food Research and Data Analysis,* Applied Science, London, 1982.
5. **Kowalski, B. R., Ed.,** *Chemometrics: Theory and Application,* (American Chemical Society Symp. Ser. 52), Washington, D.C., 1977.

Chapter 11

FACTOR ANALYSIS

I. INTRODUCTION

Factor analysis is a statistical means of reducing the redundancy in a large data set by organizing the data in terms of shared variability among the measures. For example, a food or beverage product may produce numerous sensations. These sensations may be subtended by hundreds of chemical components and will often be describable by dozens of consumer terms only some of which have clear meanings. Finally, in a product category, there may be numerous products which all seem different from one another on several sensory attributes and which are also chemically different. The many sensory attributes and chemical compounds that constitute a product are virtually impossible to make sense of when one attempts to understand the fundamental structure of a product category consisting of many product examples. It would be much easier to make sense of the category if a smaller number of underlying variables which are common to the products in the category could be identified and the products were described in terms of this smaller number of variables. Factor analysis provides this function by searching for patterns of correlations among measures taken of products. This chapter will describe the rationale behind factor analysis and present an example of its use. The purpose of this chapter is to provide the reader with a general understanding of the mechanics of factor analysis and its interpretation. No attempt is made to supply an in-depth picture which would be beyond the scope of this book. The knowledgeable reader will therefore be likely to notice where tradeoffs have sometimes been made between exact accuracy and clarity with the latter being favored. Because of the general nature of the presentation, specific citations are not used throughout the text. However, several sources have been drawn on heavily,[1-7] and the reader who is interested in more detail should consult these. The references chosen cover a range from heavily mathematical to intuitive orientation, and thus should satisfy both needs.

II. VARIANCE AND COVARIANCE

If sensory ratings or chemical determinations were made on a series of products, it would be found that each product received a different rating of each sensory attribute and had a different amount of each chemical present. Some of the products may be very close, and others may be far apart, but in general they will tend to have different sensory and chemical scores. The systematic differences among the products on a given sensory attribute or chemical compound constitute the variance of that attribute or compound. If one were to carefully search the data set, it would be eventually noticed that some attributes or compounds vary in a similar fashion. For example, it may be noticed that the higher a tea is rated on bitterness, the higher it is rated on astringency. If that were the case, bitterness and astringency would be said to share variance, or covary. This is another way of saying they are correlated. This idea is shown schematically in Figure 1. Each circle represents one variable (e.g., a sensory attribute or chemical compound), and the area of the circle represents the variance of that variable. Overlap of the circles represents a correlation or covariance between the variables, with the degree of overlap representing the degree of correlation. In a large study of many products and attributes or chemicals, it is often found that many variables will be intercorrelated in such a fashion, and several groups of intercorrelations are evident. Conceptually, the area of the circles in Figure 1 which is overlapped by all the variables in that group is a factor

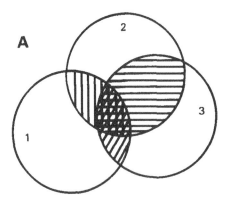

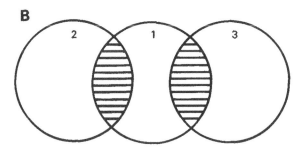

FIGURE 1. Schematic diagram of the concept of shared variability. The total area of each circle represents the total variability of a single variable. The overlapping areas represent shared variance, or covariance. In (A) the areas of the three circles which overlap represents variability shared by all three of the variables. The area of overlap is analogous to the shared variability that common factor analysis attempts to identify. In (B) variables 2 and 3 covary with variable 1 but not with each other; they share different portions of variance with 1. In a factor analysis variable 1 would be separated into two factors, one of which also contained variable 2 and the other of which contained variable 3. (From Bieber, S. L. and Smith, D. V., *Chem. Senses*, 11, 19, 1986. With permission.)

and represents a trend in the data which is common to the variables. This factor can be used to simplify understanding because the factor analysis will provide a quantitative estimate of how much each variable is representative of the factor by producing a correlation coefficient describing the degree of relationship between the variable and the factor. This correlation between the variable and factor is called a factor loading. The product of a factor analysis is a set of such factor loadings which indicate the extent to which each variable is related to the factors discovered. Since the number of factors is far less than the number of variables which load on them, the factors are much more conveniently used to describe the data than are the original variables.

Table 1 extends this conceptual reasoning into a simplified quantitative example of how trends in the data are identified as factors and how sample products can then be explained in terms of factors instead of the original variables. In the hypothetical data set shown in this table, ten different measurements have been made on each of three different product samples. The heights of the bars to the right of the sample numbers indicate the amount of each variable

TABLE 1
Factor Analytic Modeling

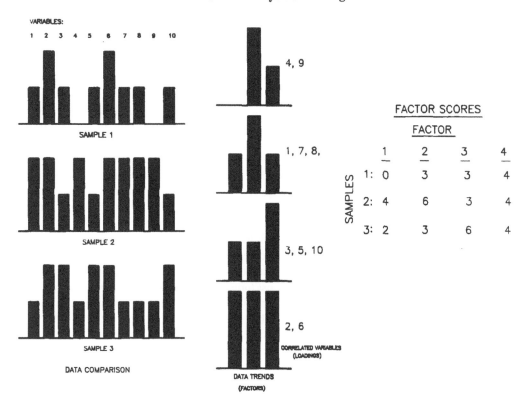

present in each sample. Initially, the data, represented by the lengths of the lines, seem to have no consistent pattern even in this very simple data set. However, closer examination reveals some trends. For example, variables 4 and 9 are correlated: they are both absent from sample 1, present in sample 2 in a high amount, and present in sample 3 in a smaller amount. Variables 1, 7, and 8 are also correlated, all being present in a high amount in sample 2, and to a lesser extent in samples 1 and 3. Further examination shows that variables 3, 5, and 10 and 2 and 6 are correlated. On this very simple level, four groups of intercorrelated variables, or factors, have thus been identified. Instead of dealing with each of the ten variables, the samples can now be described in terms of the set of four factors just identified. This is done by determining how each sample "scores" on each of the factors. In the example in Table 2 this was done by assigning a value of zero to the absence of a variable, a value of one to a low level of the variable (short line), and a value of two to a high level of the variable (tall line). In this case, factor 1 consists of variables 4 and 9. Sample 1 has neither of these variables and therefore scores zero on factor 1. Factor 2 consists of variables 1, 7, and 8. Sample 1 has a low level of each of these variables and therefore, by adding up the three low scores, has a factor score of three on factor 2. The same procedure is used to find the factor scores of each sample on each factor. The result is a set of four factor scores for each of the three samples instead of ten variable measures for each. Although this is a highly simplified example, it is analogous to how factor analysis actually works, as will be seen later in the chapter with a worked factor analysis.

The above examples illustrate that the goal of factor analysis is to identify trends or variations in the data which are shared commonly by subsets of the original variables. Another way of saying this is that factor analysis partitions variables into common and unique sources of variability:

$$\sigma = c + u + e \tag{1}$$

This model states that each variable, or the variance of a variable, σ, has three components. One component is its common variability, here designated as c. This is the variability it shares with other variables in the set and is represented in Figure 1 by the areas of the circles which overlap. Unique variability, designated by u, is that portion of the variability of the measure which is specific to it and not shared by any other variable in the set. An intuitive example of common and unique variability might be a sensory evaluation scale rating of peppery flavor. Ratings of peppery are likely to be correlated with other ratings such as hot, tingly, spicy, etc., and to the extent it correlates with these ratings it varies in common with them. However, peppery also has its own unique flavor character which, although a spice, is very different from the other terms just listed. Variability in this unique pepper flavor corresponds to the unique variability of that measurement. Error variability is the variation in measurements which is due to imprecision in instruments, techniques, and other factors which cannot be, or were not, controlled during the measurement session. It is impossible to separate true unique variability from error variability and the two are usually just lumped together and considered as one. In general, unique variability is defined as the variability left over after a factor analysis has identified the common variability. However, while factor analysis concerns itself with the common variability among a set of variables, other similar procedures do not concentrate on only this portion of the variability. For example, principle component analysis (PCA) deals with the total variability of the variables, that is, both the common and unique portions are analyzed. The end result is a set of components, which are interpreted in a fashion similar to factors, that summarize correlations among the variables in the data set. PCA is popular in chemometric applications because it is not unusual to find a chemical compound, or class of compounds, that contributes a high degree of unique variability to a data set. Since common factor analysis only concerns itself with variability that is common to all variables in the set, that important unique variability would not be detected. This same situation can also occur with sensory ratings of complex sensations. Therefore, as a practical point of thoroughness, PCA is often the method of choice.

III. THE FACTOR ANALYSIS PROCEDURE

The value of factor analysis and PCA lies in its ability to parsimoniously summarize data sets which consist of many variables and are therefore unwieldy. Although individual sensory attributes should not be ignored when interpreting sensory data, it is often more convenient to think in terms of a smaller subset of factors or components rather than a large group of variables which are often intercorrelated anyway. Consider the example of orange flavored tea. The left-most column of Table 2 lists 18 sensory attributes that can be differentiated in sample teas. While not all attributes are present in all teas, when the category is considered as a whole, each attribute will be more or less represented in the different product instances. In addition, many investigators will not pass up the opportunity to collect early data on acceptability of the products by including a rating of overall acceptability for each product. (If this is done it must be realized that Pearson's r, which is used in factor analysis, is appropriate for linear relationships between variables, and acceptability is often not linearly related to its determining variables. However, including acceptance ratings will not necessarily hurt as long as the results are not overinterpreted). A total of 19 different ratings can therefore be obtained on each product the investigator is interested in studying. Factor analytic methods allow this number to be reduced to a smaller group of factors each of which contains related attributes.

A. INPUT TO THE ANALYSIS

All factor analytic techniques search for patterns of correlations among the variables under study. In this case, interest is in analyzing the sensory ratings of the 19 attributes listed in Table 2. These ratings were obtained by having 26 different observers rate each of 7 different orange flavored teas on each of the attributes using a 20-point category rating scale as discussed in an Chapter 8. The ratings were averaged across observers and the averages were then used to compute Pearson's r correlations between every possible pair of attributes, that is, the attributes were correlated with each other using the products rated as the basis of the correlation. Consequently, seven pairs of data points entered into each correlation. This is not a particularly large number of data pairs, but it is representative of the limited data that is very often available to chemometric studies.

The resulting correlations are arranged in Table 2. In a correlation table (also called a matrix), each variable, in this case sensory attributes, is listed as a row (i.e., down the side of the table). The same variables are also listed across the top of the table (i.e., as columns) in the same order as they are listed for rows. The correlation between the variable in the first row and the variable in the first column is the correlation of a variable with itself and is therefore 1.0. The same is true for the correlation of the variable in the second row and column, the third row and column, etc. Each of these correlations is 1.0 and they constitute the main diagonal of the correlation matrix. Since the correlations in the upper right of a correlation table are the mirror image of those in the lower left of the table, sometimes only correlations below and to the left of the main diagonal are used for purposes of exposition. The correlation matrix is the starting point input to a factor analysis.

Examination of the correlation matrix can reveal some patterns in the data even at this stage. For example, sensory ratings of orange and fruit flavors and sweetness are all correlated, suggesting a fairly high degree of covariance among them. Ratings of clove and cinnamon aroma and flavor are also correlated with each other, but each of their correlations with sweet and orange ratings is lower than the correlations with each other. This suggests that clove and cinnamon flavors form a separate group of attributes that accounts for different variability than the orange/sweet group. This sort of casual scrutiny of a correlation matrix can begin to provide clues as to what factors will eventually serve to summarize a data set, but it cannot yield more than a very rough approximation because of the complexity of relationships among the variables: every rating is correlated to some extent with every other rating, and disentangling these relations requires mathematical procedures.

B. FACTORING A CORRELATION MATRIX

There are numerous methods available for factoring a correlation matrix. Although the methods differ in their mathematical details, they all share the same objective of explaining the maximum amount of observed variability with the fewest number of factors possible. Discussion of the various factoring methods is beyond the scope of this book, but a description of one method, the centroid method, will provide the general idea of the process and allow understanding of the terminology that is used in conjunction with factor analysis. The centroid method was chosen for description because it is computationally simpler than other methods, is easier to understand, and it is similar to the principal axes method which is more complicated but is frequently used in PCA. Since the methods are similar in their logic but the

TABLE 2
Correlations among Orange Tea Sensory Attributes

Attributes	1	2	3	4	5	6	7	8	9	10	11	12	13	14	15	16	17	18	19
Orange aroma (1)	1.00	-0.32	-0.16	-0.48	0.87	0.25	-0.36	-0.32	-0.57	0.91	0.87	0.83	-0.14	0.17	-0.85	-0.54	-0.29	0.57	0.34
Clove aroma (2)	0.32	1.00	0.85	-0.55	-0.21	-0.36	0.22	0.13	-0.26	-0.44	-.08	-0.35	0.98	0.85	0.37	-0.34	0.46	0.01	0.20
Cinnamon aroma (3)	-0.16	0.85	1.00	-0.50	-0.15	-0.11	0.45	0.33	-0.10	-0.38	0.16	-0.24	0.90	0.82	0.37	-0.22	0.63	0.22	0.01
Other aroma (4)	-0.48	-0.55	-0.50	1.00	-0.53	0.24	0.43	0.44	0.90	-0.33	-0.83	-0.28	-0.64	-0.76	0.36	0.94	0.17	-0.46	-0.59
Sweet taste (5)	0.87	-0.21	-0.15	-0.53	1.00	-0.10	-0.60	-0.60	-0.61	0.86	0.87	0.86	-0.10	0.12	-0.88	-0.59	-0.52	0.36	0.72
Sour taste (6)	0.25	-0.36	-0.11	0.24	-0.10	1.00	0.47	0.67	0.17	0.34	-0.05	-0.20	-0.26	-0.17	0.21	0.34	0.25	0.72	-0.46
Bitter taste (7)	-0.36	0.22	0.45	0.43	-0.60	0.47	1.00	0.96	0.69	-0.45	-0.51	-0.43	0.25	0.18	0.62	0.66	0.92	0.07	-0.73
Astringency (8)	-0.32	0.13	0.33	0.44	-0.60	0.67	0.96	1.00	0.62	-0.33	-0.51	-0.52	0.16	0.09	0.66	0.66	0.82	0.25	-0.73
Tea taste (9)	-0.57	-0.26	-0.10	0.90	-0.61	0.17	0.69	0.62	1.00	-0.54	-0.83	-0.33	-0.32	-0.48	0.52	0.49	0.49	-0.46	-0.67
Candy-like orange (10)	0.91	-0.44	-0.38	-0.33	0.86	0.34	-0.45	-0.33	-0.54	1.00	0.75	0.70	-0.31	-0.06	-0.77	-0.40	-0.49	0.60	0.47
Fresh orange (11)	0.87	-.08	0.16	-0.83	0.87	-0.05	-0.51	-0.51	-0.83	0.75	1.00	0.71	0.23	0.48	-0.76	-0.85	-0.33	0.55	.60
Other fruit taste (12)	0.83	-0.35	-0.24	-0.28	0.86	-0.20	-0.43	-0.52	-0.33	0.70	0.71	1.00	-0.22	0.05	-0.96	-0.41	-0.30	0.04	0.38
Clove taste (13)	-0.14	0.98	0.90	-0.64	-0.10	-0.26	0.25	0.16	-0.32	-0.31	0.23	-0.22	1.00	0.94	0.26	-0.42	0.51	0.15	0.18
Cinnamon taste (14)	0.17	0.85	0.82	-0.76	0.12	-0.17	0.18	0.09	-0.48	-0.06	0.48	0.05	0.94	1.00	-0.02	-0.58	0.48	0.29	0.18
Spicy taste (15)	-0.85	0.37	0.37	0.36	-0.88	0.21	0.62	0.66	0.52	-0.77	-0.76	-0.96	0.26	-0.02	1.00	0.55	0.48	-0.08	-0.48
Grassy taste (16)	-0.54	-0.34	-0.22	0.94	-0.59	0.34	0.66	0.66	0.55	-0.40	-0.85	-0.41	-0.42	-0.58	0.55	1.00	0.39	-0.31	-0.61
Woody taste (17)	-0.29	0.46	0.63	0.17	-0.52	0.25	0.92	0.82	0.49	-0.49	-0.33	-0.30	0.51	0.48	0.48	0.39	1.00	0.00	-0.65
Other taste (18)	0.57	0.01	0.22	-0.46	0.36	0.72	0.07	0.25	-0.46	0.60	0.55	0.04	0.15	0.29	-0.08	0.00	0.00	1.00	0.12
Overall acceptability (19)	0.39	0.20	0.01	-0.59	0.72	-0.46	-0.73	-0.73	-0.67	0.47	.60	0.38	0.18	0.18	-0.48	-0.61	-0.65	0.12	1.00
Column sums	1.83	2.32	3.88	-1.47	0.77	2.95	3.84	3.78	0.18	1.13	1.63	0.33	3.15	3.48	0.60	0.23	4.02	3.64	-0.72
Reflected column sums	6.65	3.02	3.06	9.25	8.19	2.83	6.28	6.74	9.18	6.13	9.77	6.43	3.71	5.64	7.24	9.55	3.72	3.94	9.12
First centroid loadings	0.61	0.28	0.28	0.84	0.75	0.26	0.57	0.61	0.84	0.56	0.89	0.59	0.34	0.51	0.66	0.87	0.34	0.36	0.83

(Total of reflected column sums: 120.45)

centroid method is simpler, it serves as an excellent vehicle to describe the factor analytic procedure.

The starting point of the centroid method, as in all other methods, is the correlation matrix as shown in Table 2. The first step is to sum the correlations in each column. In this case, there are 19 column sums, 1 for each variable. When doing this, the entire matrix must be used; that is, both the correlations above and below the main diagonal as well as the diagonal elements are used. If this is not done, the number of variables included in each sum decreases across the matrix until the final sum is only 1.0, the correlation of variable 19 with itself. (For now, the diagonal elements in the matrix used will be 1.0, but this is not always the case as will be discussed in conjunction with the issue of selecting communalities.) After each column is summed, the sum of the column sums is obtained. All the elements needed for determining the loading of each variable on the first factor are now in hand. These are obtained by dividing each column sum by the square root of the sum of the column sums. Therefore, the loading of variable 1 on factor 1 is obtained by dividing the sum of column 1 by the square root of the sum of all the column sums. This same simple division is done for each column and the first set of loadings is obtained.

1. Reflections

In all methods of factoring, the objective is to explain as much variability as possible with each successive factor. This means ensuring that the factor loadings are always as large as they can be. With the centroid method this is accomplished with the first factor by ensuring that the column sums are all as large as possible. Since the column sums are divided by the square root of the total of the column sums to obtain loadings, it is obvious that maximum loadings are obtained by maximizing column sums. While there are computational procedures available which accomplish this efficiently,[7] it can be done easily by inspecting the column sums and locating any negative sums. For example, in Table 2, columns 4 and 19 sum to negative values. After locating the negative sums, the signs of all the correlations (except for the main diagonal) involving the variable whose column sum is negative are changed, or reflected, to the opposite sign, that is, all positive correlations are made negative and vice versa. In this case the signs of all correlations in rows and columns 4 and 19 are changed to their opposite value. Recall that it is the magnitude of correlation that determines its usefulness, not its sign. Reflecting therefore does not affect the usefulness of the data. After this first reflection, the columns are again summed and examined for negative values. Variables associated with any column sums which are negative are again reflected. The process is repeated until all column sums are positive or until the sums cannot be made larger. In Table 2, this eventually leads to the reflections of variables 9, 15, 16, 7, 8, 19 (again), 17, and 6, in addition to the original reflection of variables 4 and 19. The column sums were then all positive and the process was stopped. The reflected column sums in Table 2 were then used to compute the loadings on the first factor as described, that is, each reflected column sum was divided by the square root of the total of the reflected column sums. The loadings are listed as the first centroid loadings in Table 2.

C. MEANING OF THE LOADINGS

Examining the formula for obtaining these loadings provides information about their properties:

$$F_{j1} = \Sigma r_{ij} \sqrt{\Sigma r_{ij}} \tag{2}$$

In this formula, F_{j1} is the loading of a given variable here designated as j, on factor 1. r_{ij} is the sum of the individual correlations, each of which is designated as i, of that variable with each other variable; this is the column sum. $\sqrt{\Sigma r_{ij}}$ is the square root of the sum of the column sums. The first thing evident is that this formula is an average. The centroid method therefore extracts the first factor as an average of the variables. Thus, each variable can be expected to load fairly highly on it. Second, it can be seen from the formula that variables which correlate highly and in the same direction (positive or negative) with many of the other variables will have relatively high loadings. This is because the column sums of those variables will be high. To summarize, the first centroid factor is an average factor and variables which correlate most highly with all other variables will load on it to the greatest extent.

1. Residuals and the Second Factor

If all the variability represented in a correlation matrix could be explained by one underlying factor, then there would be no further factors to extract from the matrix. To determine if this is the case, a matrix of residuals is obtained. Residuals are obtained by taking each correlation in the original correlation matrix and subtracting from it that portion of the correlation accounted for by the first factor. If the correlation between two variables is accounted for completely by one factor, then that correlation is equal to the product of the factor loadings on that one factor. Subtracting that product from the original correlation will reveal if there is a meaningful amount of variability remaining. For example, the correlation between orange aroma and clove aroma in Table 2 is –0.32. Orange aroma loads 0.61 on the first centroid and clove aroma loads 0.28 on the first centroid. Therefore:

$$0.61 \times 0.28 = 0.17 \tag{3}$$

To determine if there is still variability in the correlation between orange aroma and clove aroma that is not accounted for by the first centroid, the product of the loadings is subtracted from the correlation:

$$-0.32 - 0.17 = -0.49 \tag{4}$$

When all such products of loadings are obtained and subtracted from the appropriate correlation, the resulting correlation is called the residual matrix. If the residual matrix has values greater than zero, within error, then a second factor must be extracted.

The second factor is obtained by factoring the first residual matrix. However, with the centroid method this causes a problem. Recall that the first centroid factor is an *average* factor. Therefore, subtracting the variability it accounts for from the original matrix results in half of the elements being of positive sign and half being of negative sign. Summing each column of the residual matrix would therefore result in sums of zero. This is avoided by reflecting the elements of the residual matrix. Recall again that both positive and negative correlations of the same magnitude are equally useful, so changing the sign does nothing to the value of the data. However, if a given variable is reflected, the investigator may wish to change the sign of its factor loading once it is obtained to facilitate interpretation and communication of the results.

Once the first residual matrix is obtained and reflected, the second factor is extracted from it in the same way the first factor was extracted from the original correlation matrix. The entire procedure is repeated for successive residual matrices until the residuals equal zero or some other criteria dictate that the proper number of factors are obtained.

D. ORTHOGONAL FACTORS

The elements of the first residual matrix consist of correlations which have had the portion of their variability that is accounted for by the first factor removed. Consequently, when the second factor is extracted from the first residual matrix, the variability accounted for by the first factor is not considered in the computations. Statistically, this means that the second factor is independent of the first or, using the statistical term, it is orthogonal to it. In addition to being orthogonal to the first factor, successive factors obtained by the centroid method will all be bipolar, i.e., they will have both positive and negative loadings. This is because the first factor is an average, and of necessity, a subsequent factor which is independent of it must have both positive and negative loadings around the average. With a little thought, it will be realized that this is true for each factor obtained. They are all independent of one another and all will be bipolar. The first centroid, because it is an average, will always be a general factor with many variables loading substantially on it.

IV. PRINCIPAL AXES METHOD

The centroid method as described above provides an understanding of the executional operations required to perform a factor analysis. A very commonly used and similar procedure is referred to as the principal axes method. This method is frequently used in PCA and will be discussed in a worked example here. The major difference from the centroid method is that the principal axes method weights the variables in such a way as to ensure that the first factor explains the maximum amount of variance possible. Residual matrices are obtained as with the centroid method, but the the variables are again weighted so that the second factor explains the maximum amount of remaining variability possible, and so on. If reflections are carried out in the centroid method so that the column sums of the residual matrices are maximized, the two methods give very similar results. The advantage of principal axes is that it explains more variance with each factor and this becomes more obvious as more factors are needed to describe the data set. In addition, it is possible to obtain slightly different results from the centroid method of extraction with the same data set because methods and details of reflection can differ. However, principal axes result in only one solution for a given data set. The disadvantage of principal axes are that it is computationally more complex and can be carried out only by computer (as opposed to the centroid method which can be done on a hand calculator). As a result, the investigator is likely to see only the end result of principal axes and will not have a feel of how the intermediate steps evolved. The practical matter is that principal axes is probably the method of choice in current practice and is readily available in computer statistical packages. Despite its differences from the centroid method, it is approaching the same goal with the same logic.

V. AN EXAMPLE OF PRINCIPAL COMPONENTS ANALYSIS

The correlations in Table 2 were subjected to a principal components analysis (PCA). Recall that a PCA involves both common and unique variance associated with each variable. This is accomplished by having values of 1.0 as each entry in the main diagonal of the correlation matrix. While it may seem obvious that this should always be the case since the correlation of a variable with itself is in fact 1.0, other types of analyses, such as common factor analysis, require different values in the diagonal. The factoring technique used on the data set was principal axes. The initial results of the analysis are a table of unrotated factor loadings. These are shown in Table 3. The variable names are listed in the left column of this table, and the numeric entries are the correlations between each of the variables and each of the factors. These are the factor loadings and are interpreted in exactly the same manner as any other correlation. For example, the correlation between orange aroma and the first factor

TABLE 3
Table of Unrotated Factor Loadings

Attribute	F_1	F_2	F_3	F_4	F_5	F_6	h^2
Orange aroma	0.81	–0.12	0.54	0.19	–0.81	–0.01	1.00
Clove aroma	–0.08	0.95	–0.24	–0.01	0.11	0.14	1.00
Cinnamon aroma	–0.14	0.93	0.08	0.12	0.15	–0.28	1.00
Other aroma	–0.72	–0.66	0.01	0.15	0.11	0.08	1.00
Sweet taste	0.92	–0.12	0.18	0.19	0.27	–0.05	1.00
Sour taste	–0.23	–0.18	0.89	–0.35	–0.03	0.03	1.00
Bitter taste	–0.78	0.28	0.49	0.25	0.08	0.00	1.00
Astringency	–0.76	0.20	0.60	0.02	0.07	0.08	1.00
Tea taste	–0.85	–0.34	0.05	0.34	0.19	–0.10	1.00
Candy–like orange	0.78	–0.31	0.50	–0.05	0.15	0.15	1.00
Fresh orange taste	0.92	0.25	0.26	0.07	–0.04	–0.10	1.00
Other fruit taste	0.74	–0.25	0.16	0.60	0.00	–0.04	1.00
Clove taste	0.02	0.99	–0.08	0.06	0.07	0.09	1.00
Cinnamon taste	0.23	0.94	0.11	0.17	–0.09	0.12	1.00
Spicy taste	–0.85	0.25	–0.08	–0.41	0.10	–0.09	1.00
Grassy taste	–0.83	–0.43	0.14	0.13	0.28	0.02	1.00
Woody taste	–0.63	0.53	0.38	0.41	–0.07	0.06	1.00
Other taste	0.36	0.24	0.78	–0.44	0.08	–0.08	1.00
Overall preference	0.77	0.12	–0.35	–0.15	0.49	0.08	1.00
Eigenvalue	8.6	5.2	3.0	1.4	0.6	0.2	
Eigenvalue/19	0.45	0.27	0.16	0.07	0.03	0.01	
Cumulative variance accounted for (%)	45.3	72.7	88.7	96.0	98.9	100.0	

is 0.81. Squaring this value results in a coefficient of determination of 0.65, meaning that 65% of the variability in orange aroma can be accounted for by factor 1. By comparing the first principal component from Table 3 to the first centroid in Table 2 it can be seen that the absolute values of the loadings are similar, i.e., variables that load highly on the first centroid also load highly on the first principal component.

Several points can be made about the table of unrotated factor loadings. First, it can be seen that the loadings, on the whole, are largest between each variable and the first factor. After that, the loadings tend to decrease until loadings on the last factor are all very small. This is a result of the principal axes technique maximizing the variability explained by each successive factor. The amount of variability explained by each factor can be computed by summing up the coefficients of determination computed from the correlations between each variable and the factors. For example, by squaring each loading on the first factor and then adding them all up, a value of 8.6 is obtained. Such values are called eigenvalues and are listed under each factor in Table 3. The amount of variability accounted for by a factor is obtained by dividing the eigenvalue for the factor by the total number of variables in the table and then multiplying by 100. For factor 1, this value is 45.3 = (8.6/19 × 100). The reasoning here is that the total variance in a correlation matrix is equal to the number of variables it contains because computing correlations standardizes variables to have an average of 0.0 and a variance of 1.0. In this case, the total variance is equal to 19. Therefore, the amount of variability accounted for by a factor can be divided by the number of variables in the matrix to result in the proportion of variability explained by the factor. Multiplying that proportion by 100 results in the percent of variability accounted for by the factor. Inspection of the eigenvalues and the corresponding percentage of variability accounted for by each factor shows successively smaller amounts accounted for by each factor.

When the eigenvalues are summed, the obtained value is 19. This is the same as the number of variables in the correlation matrix. Therefore, this analysis accounted for 100% (19/19 × 100) of the variability in the original matrix. This is expected, in fact it is required, because a PCA involves both common and unique variability of each variable. In a common factor

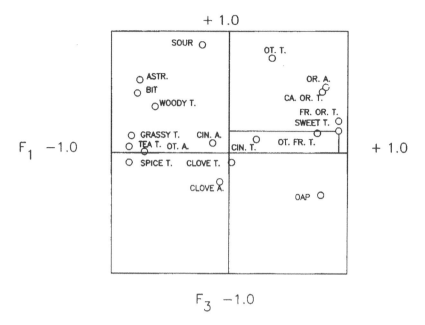

FIGURE 2. Pattern of unrotated loadings on factors 1 and 3 from Table 3.

analysis, which concentrates only on common variability, the total variability explained by the factors would be less than 100% because unique variability of each variable would not be accounted for.

The proportion of variability of each variable accounted for by the obtained factors is determined by squaring the values in the row of loadings corresponding to the variable of interest and then adding them. For example, the percentage of variability in orange aroma ratings accounted for by the factors is $0.81^2 + (-0.12^2) + 0.54^2 + 0.19^2 + (-0.08^2) + (-0.01^2)$ = 1.0. Multiplying this value by 100 results in the percentage of variability of each variable accounted for by the factors. As with the eigenvalues, it can be seen that the coefficient of determination is the basis of this calculation. In PCA, since both common and unique variability are accounted for, these values will all equal 1.0. In a common factor analysis, which accounts for only common variability, the values are computed the same way and are called *communalities*. In that case, the values represent how much each variable has in common with the set of factors and unique variability is ignored.

More detailed inspection of Table 3 reveals some peculiarities. For example, it can be seen that most variables load highly on the first factor. While this is good in that it means the first factor accounts for a large amount of variability, it does make interpretation of the factor difficult. What common underlying variable could account for a good part of every correlation in the original data set? While it may be tempting to describe this factor, or underlying variable, as a general flavor factor, this is not very comfortable. It seems intuitively messy to include basic tastes like sour and sweet, tactile sensations like astringency, and aromas of various fruits and spices on the same continuum. The reason for this peculiar set of affairs is that the factors have not yet been rotated.

VI. FACTOR ROTATION

Figure 2 is a geometric representation of the loadings of each of the 19 attributes in Table 3 on factors 1 and 3. Since correlations range from −1.0 to +1.0, both axes also cover the same range. The horizontal axis in this figure represents factor 1, and the vertical axis represents factor 3. Computation of Pearson's correlation coefficient standardizes scores, so the variables

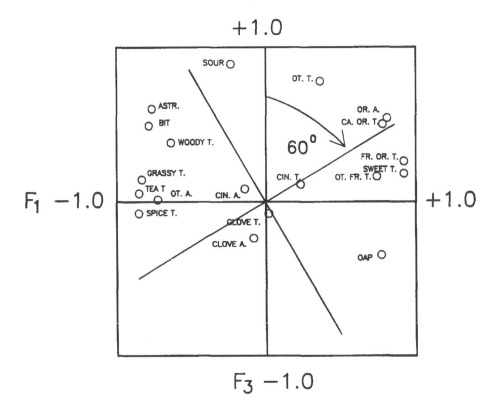

FIGURE 3. Mechanical rotation of axes to maximize loadings on factors 1 and 3. The plotted points are the same as in Figure 2 and Table 3. The arrow in the upper right quadrant shows the clockwise direction of rotation of the factors about 60° to place the axes in closer proximity to the variables.

can be represented as having the same origin at 0.0. Finally, recall that each factor extracted is independent of the others, and this situation is represented by right angles between the axes. In a geometric representation of factor loadings, the loading is represented by the projection of perpendicular lines from the point corresponding to the variable of interest to each of the axes in the plot. The point at which the line intersects each of the axes is the correlation between the variable and the factor, i.e., the factor loading. In Figure 2, perpendicular lines have been extended from the point corresponding to the attribute sweet taste to both axes. The lines intersects the X and Y axes at 0.92 and 0.18, respectively. In Table 3 it can be seen that these are the factor loadings of sweet taste on factors 1 and 3. The entire graph was constructed from Table 3 by using the loadings of factors 1 and 3 as X and Y coordinates, respectively, of this two-dimensional plot. The same procedure can be carried out with any set of two or three factors to provide a visual impression of how the variables group together.

Examination of Figure 2 shows visually what it means when it is said that the first factor is extracted so that loadings on it are maximized. If a perpendicular line were extended from each variable to both axes, it would be seen that almost all of the variables have a higher projection onto factor 1 than on factor 3, and the same thing would be found for any other factor compared to factor 1. This is the mathematical consequence of applying weights to variables in such a way as to maximize the amount of variability each factor explains. Factor rotation can be thought of adjusting those weights so that the same amount of variability originally extracted from the entire analysis is redistributed among the factors to facilitate interpretation.

Rotation can be done mechanically or mathematically. Mechanical rotation is a very simple process when only two or three factors are obtained. The procedure is illustrated in Figure 3.

TABLE 4
Procedure for Mathematical Factor Rotation

1. Unrotated factor loadings of *sweet taste* on factors 1 and 3 from Table 2:

	Factor 1 (F_1):	Factor 3 (F_3):
Sweet taste	0.92	0.18

2. Axes in Figure 2 are rotated 61° clockwise as shown in Figure 3.
   angle of rotation of F_1' with F_1 = 61°; cos 61° = 0.48
   angle of rotation of F_1' with F_3 = 151°; cos 151° = -0.87
   angle of rotation of F_3' with F_1 = 29°; cos 29° = 0.87
   angle of rotation of F_3' with F_3 = 61°; cos 61° = 0.48
   (F_1' and F_3' are rotated factors 1 and 3, respectively)

3. Loadings on F_1' = 0.48(0.92) - 0.87(0.18) = 0.28
   Loadings on F_3' = 0.87(0.92) + 0.48(0.18) = 0.89

4. Factor loadings of sweet taste on rotated factors 1 and 3:

	Factor 1 (F_1'):	Factor 3 (F_3'):
Sweet taste	0.28	0.89

5. Sum of cross products of weights equals 0.0 to maintain orthogonality of factors:
   weights for F_1' + weights for F_3' = 0.0
   [(0.48)(0.87)] + [(-0.87)(0.48)] = 0.0

6. Sum of squared weights equals 1.0 to maintain unit length of factor axes:
   0.48^2 + -0.87^2 = 1.0 (within rounding)

There, the axes of the graph in Figure 2 have been rotated about 60° clockwise. This value was selected because visual inspection of the graph seemed to indicate that rotating the axes that amount would increase the projections of the variables in the upper right quadrant onto factor 3. In a similar fashion, the cluster of variables that initially had high negative projections on factor 1 would now have high negative loadings on factor 3. Finally, the group of three variables in the upper left quadrant now loads more highly on factor 1 than they did initially. The new loadings on factors 1 and 3 can be determined directly from Figure 3. This is done very easily by extending perpendicular lines from each point to the rotated axis and reading the values of the points on the axes at which the lines intersect it. Those values are the rotated factor loadings.

The end result of rotating these two factors is apparent in the visual comparison of Figures 2 and 3, where it can be seen that the number of variables loading highly on factor 1 was decreased, and the number loading on factor 3 was increased. Importantly, the relationships among the variables themselves did not change. Figure 3 shows that it is the axes which move relative to the variables. The clusters of variables and relationships among them remained constant. Rotating the axes only redistributed the weighting of the variables so as to make them nearer to (or farther away from) each of the axes rotated. Doing this makes it easier to visually locate clusters along the axes.

If only two or three factors are found, then rotation can be done mechanically exactly as shown here. With three axes, the investigator may attempt to rotate all three at once on three-dimensional plots or use graphic computer software to accomplish the task. The only rule is that all three axes must remain at 90° angles to each other to maintain independence of the factors. Alternatively, axes could be rotated two at at time until all pairs are rotated. Obviously, this procedure will quickly become quite laborious as the number of factors increases. In that case mathematical methods are used.

The gist of the computational procedure for rotating factors mathematically is given in Table 4. The basic idea is to weight the factors so as to maximize the loadings of variables which seem to cluster together on one factor, and minimize the loadings of those variables on other factors. In Figure 3 it can be seen visually how this is done. While loadings on the rotated factors can be read directly from the figure as described earlier, the mathematical procedure allows more exact determination. Mathematical determination is required if several factors are being rotated and is practically preferred in all cases.

Table 4 uses the data in Figure 3 to exemplify the procedure. The first step is to determine the angle of rotation between the rotated axes and the original unrotated axes. In Figure 3 the angle between the rotated axis F_1' and the original unrotated axis F_1 is 61°. This is also done for the angle formed between F_1 and F_3, and for the angles between F_3' and the original axes as shown in Table 4. Geometrically, the correlation between any two variables, or between any variable and a factor, is represented by angle between them when plotted as in Figures 2 and 3. This can be understood intuitively when it is realized that the cosine of an angle of 0° is 1.0, that is, if two points are in exactly the same location on the plot, there will be no angle between them and correlation will be 1.0. Conversely, if two points are on exactly opposite points of the graph, the angle between them is 180° and the cosine of that angle is –1.0; the correlation between the variables –1.0. Therefore, in order to increase the absolute value of a correlation between a variable and a factor, the angle between them is either minimized or maximized. This is done mathematically by multiplying the loading of the variable on the unrotated factor by the cosine of the angle between the unrotated and rotated factor. Table 4 shows this using the loadings of sweet taste on the unrotated factors 1 and 3. Both of these factors are simultaneously rotated by multiplying the original loadings by the cosines of the corresponding angles between rotated and unrotated factors and summing the products; this is shown in step 3 of Table 4. Step 4 of Table 4 shows the loadings of sweet taste on the rotated factors F_1' and F_3'. Note that the loadings obtained correspond very closely to what the loadings obtained by visual estimation from Figure 3 would have been. To obtain the rest of the loadings, the same multiplications are carried out on each of the variables in turn.

Mathematical rotation can be carried out without the aid of a graph by examining the table of unrotated factor loadings and selecting weights that maximize the loadings of certain variables and minimize the loadings of others. There are certain rules that must be followed when this is done. First, the correlations range from –1.0 to +1.0 and factor axes have unit length. In order to preserve unit length, the sum of the squared weights must equal 1.0. In addition, the sum of the cross products of the weights must equal 0.0 in order to maintain the independence of the rotated factors. With this in mind, any number of factors may be rotated and rerotated until a pleasing result is obtained.

What is a pleasing result? A pleasing result is one which maximizes the loadings of a small number of different variables on each factor and minimizes the loadings of the other variables on those factors. Geometrically, this means that each axis in the plot of the factors comes as close as possible to one cluster of variables and is as far away as possible from the other clusters. In some cases, the axis comes close to one cluster at both its positive and negative extremes, which can be interpreted to mean that the variables are measuring opposite ends of a continuum which represents the underlying variable.

Rotating factors mechanically and mathematically by hand quickly becomes too laborious when more than two factors are found. Fortunately, statistical computer packages routinely provide for rotation of the factors as unrotated factors are not very interpretable. The results of the orange tea factor analysis were rotated by computer using the procedure called Varimax which provides rotated factors containing a relatively small number of variables which load highly on them, and distributes the variance explained by the factors in as even a pattern as possible. The factor table produced by this rotation is in Table 5. The first thing to be noticed here is that the loadings of *sweet taste* on rotated factors 1 and 3 are not the loadings obtained by mechanical and mathematical rotation as described earlier. The reason is that the rotation discussed in the examples above rotated only two factors, 1 and 3. The rotation performed to result in Table 5 was done simultaneously on all six of the original factors. Therefore, the axes were rotated with respect to factors 2, 4, 5, and 6, as well as 1 and 3. Consequently, different weights were applied to maintain unit length and orthogonality. It is still instructive to view

TABLE 5
Table of Rotated Factor Loadings

Attribute	F_1'	F_2'	F_3'	F_4'	F_5'	F_6'	h^2
Orange aroma	0.90	−0.02	−0.19	−0.40	−0.07	0.02	1.00
Clove aroma	−0.28	0.94	−0.04	0.13	0.12	0.08	1.00
Cinnamon aroma	−0.14	0.90	0.21	−0.06	0.04	−0.34	1.00
Other aroma	−0.33	−0.70	0.59	0.21	0.04	0.08	1.00
Sweet taste	0.86	0.00	−0.37	−0.14	0.32	−0.08	1.00
Sour taste	−0.03	−0.25	0.39	−0.87	−0.13	0.07	1.00
Bitter taste	−0.29	0.22	0.91	−0.17	−0.11	−0.03	1.00
Astringency	−0.36	0.12	0.83	−0.39	−0.09	0.07	1.00
Tea taste	−0.35	−0.38	0.80	0.27	0.05	−0.13	1.00
Candy–like orange	0.75	−0.23	−0.27	−0.50	0.20	0.16	1.00
Fresh orange	0.76	0.34	−0.46	−0.29	0.00	−0.09	1.00
Other fruit taste	0.98	−0.10	−0.10	0.16	0.01	−0.05	1.00
Clove taste	−0.13	0.99	0.00	0.02	0.06	0.04	1.00
Cinnamon taste	0.16	0.97	−0.02	−0.07	−0.11	0.09	1.00
Spicy taste	−0.92	0.14	0.33	−0.11	0.03	−0.11	1.00
Grassy taste	−0.41	−0.49	0.75	0.08	0.16	−0.01	1.00
Woody taste	−0.17	0.50	0.81	−0.02	−0.24	0.04	1.00
Other taste	0.23	0.21	−0.08	−0.94	0.05	−0.07	1.00
Overall preference	0.33	0.20	−0.67	0.07	0.63	0.00	1.00
Eigenvalue	5.44	5.14	4.95	2.52	0.70	0.22	
Eigenvalue/19	0.28	0.27	0.26	0.13	0.03	0.01	
Cumulative variance accounted for (%)	28	55	81	94	97	98	

the geometry of the rotation as shown in Figure 4. This figure ought to be compared with Figure 2 because the same variables are now plotted around the rotated factors axes F_1' and F_3'. A visual comparison will show that the same clusters of variables can be found, but are now in different positions around the axes. This exercise is facilitated by referring to the factor loadings in Tables 3 and 5 and using the loadings as coordinates to locate each variable in the plots. This was done in Figure 5, which shows the unrotated loadings on the left and the rotated loadings on the right. Each of the attributes has been located in both plots and labeled. The exact orientation of the clusters is not the same because the rotation was done in six dimensions (factors) simultaneously. Therefore, in addition to moving clockwise and counter-clockwise, the axes were moved into and above the plane of the paper as well as in higher-dimension spaces. The next thing that is evident is that there are fewer variables that load highly on F_1'. Reference to Table 5 shows that those variables which no longer load on F_1' now load highly on other rotated factors. F_3', which before rotation had only two substantial loadings (sour taste and other taste as seen in Table 3), now has six substantial loadings (bitter taste, astringent, tea taste, grassy taste, woody taste, and overall preference) as in Figure 3 and Table 5.

Another difference between the rotated and unrotated factors is in the amount of variance explained by each factor. The eigenvalues in Table 5 indicate that rotated factors 1, 2, and 3 now explain about equal amounts of variability. This more even distribution of explained variability is a consequence of the Varimax rotation procedure which evens out the variability accounted for. This should be contrasted with the eigenvalues in Table 3. Those values decrease in a consistent, negatively decelerating fashion. That is the consequence of the original selection of factoring weights which are chosen to maximize the amount of variability explained by successive factors rather than evenly distribute it. However, in both cases the total amount of variability explained by all the factors combined remains the same: rotation has only distributed it differently among factors. The sums of the squared factor loadings in

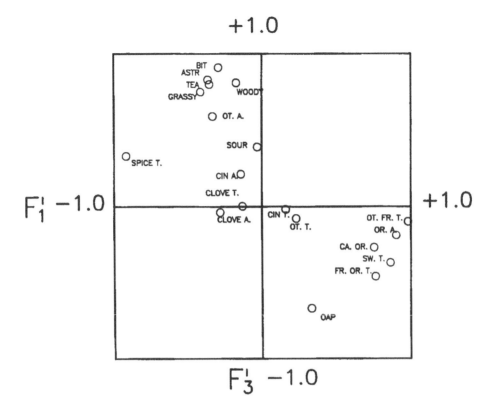

FIGURE 4. Factors 1 and 3 after Varimax rotation. The plotted values are taken from Table 5. The plotted values are not the same as those in Table 4 or in the mechanical rotation shown in Figure 3 because the Varimax rotation rotated all factors simultaneously rather than only two, as was done in the other examples.

each row, which represent the amount of variability of each variable accounted for by the factors, remains unchanged after rotation. This is again because the same variability is accounted for by the factors, but is distributed differently among them.

In summary, rotation makes interpretation of factors easier because it ensures that the factors "pass through", using the geometric analogy, clusters of variables which seem to be related by virtue of their intercorrelations. The relations among the variables do not change, and the amount of variance accounted for by the factors also does not change. It is simply redistributed in a more intelligible fashion.

VII. SELECTING THE BEST NUMBER OF FACTORS

How many factors are needed to explain a data set adequately? At this point, the judgment of the investigator begins to become very important because there are no hard and fast rules for selecting the best number of factors. There are, however, a number of guidelines that can be followed to aid judgment. One useful tool is called a *scree plot,* as is shown in Figure 6. A scree plot is a plot of the eigenvalues as a function of the factor ordering. Figure 6 shows scree plots of eigenvalues for both the unrotated and rotated factors obtained in the orange tea factor analysis. The plot of unrotated loadings exhibits the negative deceleration mentioned above and gives no indication of a clear cutoff point in the variance explained by the factors. The plot of eigenvalues from the rotated factors is quite different. The first three factors explain almost equivalent amounts of variability and a sharp drop occurs at the fourth factor. The fourth factor still explains a considerable amount of variability, 13%, but the last two

173

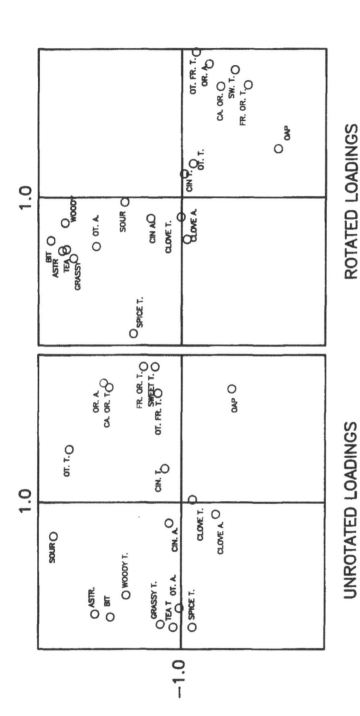

UNROTATED LOADINGS

ROTATED LOADINGS

FIGURE 5. Comparison of unrotated (left side) and rotated (right side) factors. Inspection of the two graphs shows that the variables which cluster together before rotation also cluster together after rotation. For example, the variables in the upper right of the unrotated plot appear in the lower right of the rotated plot. The orientations are changed slightly because the plots show only two factors while the rotation does not change the relationships among the measured variables, but does simplify identification of factors by maximizing loadings on one factor and reducing loadings on all others.

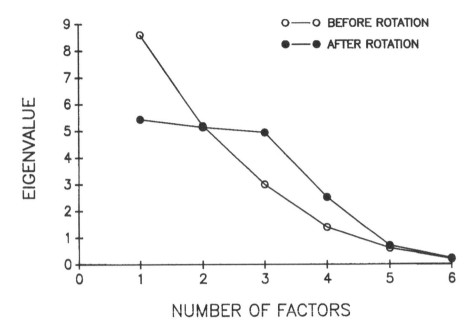

FIGURE 6. Eigenvalues plotted as a function of factor number. These plots are called scree plots. Before rotation the eigenvalues, which indicate the amount of variability accounted for by a factor, decrease as a smooth negatively decelerating function of factor number. After Varimax rotation, the eigenvalues for the early factors are more equal and decrease quickly starting with the fourth factor.

factors show another clear drop to negligible levels. This plot therefore suggests that at least three, and perhaps four, factors are useful for describing the entire data set. By adding up the percent variance explained by the first four factors, it is found that they account for about 94% of the total variability in the data set, which is very good.

Another method for determining the number of factors that ought to be considered useful is by retaining only those factors which have an eigenvalue greater than the average eigenvalue. The average eigenvalue of the rotated factors in this data set is 3.16. By this criterion, the first three factors would be retained as useful. A related criterion is based on the fact that the variance of a standardized variable is equal to 1.0. Computation of Pearson's Product Moment correlations (which are used here) standardizes the variables. Therefore, the total variability of the data set is 19, or the sum of the individual variances. The average variability accounted for by any one variable is therefore 19/19 = 1.0, that is, the total variability of 19 divided by 19 variables. It therefore makes some sense to regard any factor which explains more than the average variability accounted for by one variable as meaningful. By this criterion, four factors would be regarded as meaningful in the present analysis.

A more statistically precise method of determining whether a factor explains a significant amount of variability is by performing statistical tests on successive residual matrices to determine if the correlations they contain differ significantly from zero. If a residual matrix does contain correlations which differ significantly from zero, then the factor extracted from that matrix is retained as explaining a meaningful amount of variability. This method is precise but has a drawback; that is, when a data set is based on a large number of data points, correlations which are quite small can still be significantly different from zero. Factors based on those correlations will be hard to interpret.

The last method of determining the usefulness of a factor should be used in conjunction with one or some of the other methods discussed. The investigator should use his or her knowledge of the system being worked with to provide a judgment of whether or not the

TABLE 6
Table of Rotated Factor Loadings for Four Retained Factors

Attribute	F_1'	F_2'	F_3'	F_4'	h^2
Orange aroma	-0.90	-0.02	-0.19	-0.40	1.00
Clove aroma	-0.28	0.94	-0.04	0.13	0.98
Cinnamon aroma	-0.14	0.90	0.21	-0.06	0.87
Other aroma	-0.33	-0.70	0.59	0.21	0.99
Sweet taste	0.86	0.00	-0.37	-0.14	0.89
Sour taste	-0.03	-0.25	0.39	-0.87	0.97
Bitter taste	-0.29	0.22	0.91	-0.17	0.98
Astringency	-0.36	0.12	0.83	-0.39	0.98
Tea taste	-0.35	-0.38	0.80	0.27	0.97
Candy-like orange	0.75	-0.23	-0.27	-0.50	0.93
Fresh orange	0.76	0.34	-0.46	-0.29	0.98
Other fruit taste	0.98	-0.10	-0.10	0.16	1.00
Clove taste	-0.13	0.99	0.00	0.02	0.99
Cinnamon taste	0.16	0.97	-0.02	-0.07	0.97
Spicy taste	-0.92	0.14	0.33	-0.11	0.98
Grassy taste	-0.41	-0.49	0.75	0.08	0.97
Woody taste	-0.17	0.50	0.81	-0.02	0.93
Other taste	0.23	0.21	-0.08	-0.94	0.98
Overall preference	0.33	0.20	-0.67	0.07	0.60
Eigenvalues	5.44	5.14	5.95	2.52	
Eigenvalue/19	0.28	0.27	0.26	0.13	
Cumulative variance accounted for (%)	28	55	81	94	

factors are meaningful. In the present case, scree plots and eigenvalues suggest that three or four factors ought to considered as meaningful. Therefore, the decision is made to tentatively retain the fourth factor as meaningful on the basis of its eigenvalue and the judgment that sour taste, which loads substantially on this factor, might contribute significantly to the flavor of tea and could be related to some physical parameter of the tea, such as pH or acidity. Therefore, until the factor can be more clearly shown to be not meaningful, the conservative decision to retain it as useful will be made.

VIII. INTERPRETING THE FACTOR TABLE

Once factors have been rotated and the number of factors to consider as meaningful has been decided upon, the last remaining task is to determine what the factors describe, i.e., what is the underlying variable that results in a group of variables all being intercorrelated and emerging as a factor? To do this, the four rotated factors have been arranged into Table 6 along with their eigenvalues and the amount of variance of each of the variables they account for.

What factors represent is determined by examination of the variables which load on them. To do this, it must be decided what constitutes sufficient loading to regard a variable as representative of a factor. In earlier discussions the term "substantial" loading was used to describe some loadings. The magnitude of loading which is considered substantial is largely a matter of judgment. One often-used criteria is to regard any loading below an absolute value of 0.30 as insignificant. This criteria can be meaningful if a large data set is used. However, when smaller sets are used, the probability that a correlation will reach that value, or be greater, by chance alone increases. Therefore, as the data set decreases in size, the magnitude of loading to be taken seriously must increase. With small data sets, as in the present case, loadings which reach an absolute value of at least 0.60 can probably be considered. Variables with smaller loadings should be included for consideration only if the investigator can recognize a logical reason for their being considered.

Examination of Table 6 shows that five variables have substantial positive loadings on factor 1. These are orange aroma, sweet taste, "candy-like" orange taste, fresh orange taste, and other fruity tastes. The only substantial negative loading on factor 1 is spicy taste. This suggests that a large amount of variability in product ratings is accounted for by orange flavor and other tastes associated with it, and by spicy flavor. Further, the positive loadings of the "orange" variables in conjunction with the negative loadings of spicy tastes implies that products tended to be dominated by either one or the other: teas high in orange flavor were low in spicy taste and vice versa.

The second component of observed variability in the data set is apparently accounted for largely by specific spice flavors of clove and cinnamon, which load in excess of 0.90. These attributes, whether "tasted" in the mouth or smelled (aroma), dominated some of the products, while others had very little of these attributes and tended to be rated highly for "other" aromas (–0.70 loading) not resembling the cinnamon/clove flavors. This component accounts for just about as much variability (27%) as the first component (28%) for a total of 55% of the observed variability in product ratings being accounted for by these two components. From this it can be seen that the products evaluated were dominated by orange/fruity flavors, and spicy flavors especially clove and cinnamon.

The third component of variability is accounted for by the traditional flavors and tastes associated with tea: bitter (0.91), astringent (0.83), "tea" taste (0.80), and grassy (0.75) and woody (0.81) tastes. Finally, the last factor accepted as meaningful, factor 4, has only two substantial loadings. These are sour taste (–0.87) and other tastes (–0.94). Both of these loadings are large and negative, but even when considered together no immediate rationale for their emerging on the same factor is apparent. In such cases, examination of the pattern of other, smaller loadings can sometimes provide clarification. For example, loadings of orange aroma (–0.40), astringency (–0.39), and candy-like orange taste (–0.50) are in a range that could be considered low to moderate loadings for this size of data set. While caution is therefore dictated, these other loadings lend support to the conjecture that the factor is representing the sourness and perhaps "tingly" mouth feel characteristics associated with citrus fruit flavors. Further clarification of this factor might be forthcoming when a similar analysis of the chemical makeup of the products studied is examined.

In summary, four main dimensions, or components of variability, have emerged as accounting for the variability among the 19 attributes rated. These can be identified as variability along an orange flavor-general spicy dimension, a cinnamon-clove dimension, a traditional tea flavor dimension, and perhaps a sour dimension. These four dimensions account for a total of 94% of the observed variability in the data set. In addition, a very substantial part of the variability of each variable is accounted for by the four dimensions. This can be seen in column 6 of Table 5 which contains the sums of the squared factor loadings of each variable. The only exception is overall preference ratings where 60% of the variability is accounted for. This value could be increased to 99% by including a fifth factor (see Table 5). However, no other variable loads substantially on the fifth factor and it only accounts for a total of 3% of the variability in the data set. Therefore, that factor was not contained in the final set. It is likely that the nonlinearity of preference ratings in general resulted in that variable having a moderate positive loading on factor 3, and a moderate negative loading on factor 5. As will be shown below, these four dimensions can now be used to greatly economize the description of the seven teas that were rated.

IX. FACTOR SCORES

A factor score is the sum of the products of the factor loadings and the values of the variables on which each product was measured. For example, to obtain a factor score on factor 1 for product R, the variables, in this case sensory attributes, which load heavily on factor 1

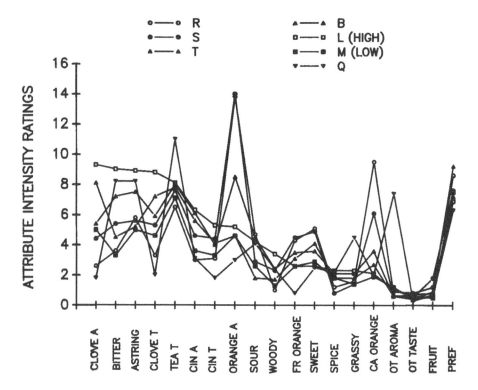

FIGURE 7. Average category ratings of each sensory attribute of the seven teas used to obtain data for the factor analysis. These are the starting point data of the factor analysis before calculating intercorrelations among the attributes.

are selected. Then, the loadings of these variables on factor 1 are multiplied by the value these variables were given in the original sensory rating session. The factor score of product R on the first factor is therefore obtained by 0.90(13.8) + 0.86(5.2) + 0.75(9.5) + 0.76(4.3) + 0.98(1.23) + −0.92(1.23). The loadings are those for orange aroma, sweet taste, candy-like orange flavor, fresh orange flavor, fruit taste, and spicy taste as found in Table 6. The numbers in the parentheses are the averages of the sensory ratings of product R that were obtained in the sensory evaluation session of the tea products studied. In this case, it was decided that the factor scores would be computed by using only those variables with relatively large loadings on the factor. The variables chosen in this way are called indicator variables and the factor score represents a weighted sum of the sensory ratings given to the product. The weights are the factor loadings. Factor scores may also be computed by summing the products of every variable and its corresponding factor loading if the investigator wishes to include relatively minor contributions to the score. In most cases it is more convenient to use only the smaller set of variables with demonstrated high loadings on the factor as indicator variables. The variables used here are the ones associated with high loadings as indicated by underlining in Table 6. Factor scores are useful because they combine the variability of several indicator variables into one composite score. Since the indicator variables load on the same factor, they are all indicative of the same continuum of variability. Thus, rather than describing a product by the several indicator variables, the use of one factor score increases efficiency and simplifies interpretation.

This is best seen in graphical display of the results of the sensory ratings of the seven teas studied in this example. Figure 7 shows the sensory attribute intensity ratings of each of the 7 teas on each of the 19 attributes for which they were rated. The attributes are arranged in order of decreasing intensity rating for the high reference tea (product L). This was done

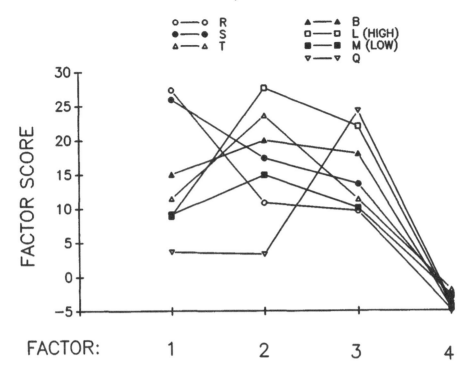

FIGURE 8. Plot of factor scores for each of the seven teas. The four factor scores represent a condensation of the 19 original variables plotted in Figure 7 into a more manageable and interpretable number.
The plot indicates that the seven teas are all virtually identical in score on the fourth factor. This factor
is therefore not of practical use in describing the teas even though some other criteria might indicate it as
statistically relevant. The seven teas separate very clearly on the other three factors.

arbitrarily in order to add clarity to the visual representation. Even with this simplification,
Figure 7 is cluttered and difficult to decipher. It is virtually impossible to follow the path of
any one product through the maze of increasing and decreasing attribute intensities and, even
if subsets of different attributes were placed on separate graphs, the decision of which ones
to graph together would be essentially arbitrary.

Figure 8 plots the factor scores obtained as discussed above for each of the seven products
on each of the four factors retained from the analysis. This figure should be contrasted with
Figure 7. Obviously the visual impression of the data is more pleasing in the sense that it is
much less cluttered and trends in the data are immediately apparent. It can be seen, for
example, that product Q scores very low on the orange flavor and spice flavor factors (factors
1 and 2, respectively), while scoring very high on factor 3, the traditional tea flavor factor.
Conversely, product R and product S are both very orange in flavor, while product L is the
most spicy of the products. It is also clear that the use of factor scores provides for better
discrimination among the products. This is evident by the greater spread of factor scores
relative to the spread of the attribute ratings seen in Figure 7. Finally, Figure 8 argues that the
controversial fourth factor is probably not going to contribute to our understanding of this sort
of product. Even though factor scores spread out the products for greater discriminability,
factor scores on the fourth factor are virtually all equal. This means that the products cannot
be differentiated on the basis of this factor and it therefore cannot help us understand
differences among them. This information on the fourth factor makes a good point on the
practical use of factor analysis. Statistically, it may be decided that the fourth factor ought to
be included, as it does seem to explain a meaningful amount of variability. While this may be
true from a statistical standpoint, in practice the factor is not useful because it does not

TABLE 7
Factor Scores for Seven Orange Tea Products

Tea product	F_1	F_2	F_3	F_4
R	27.4	10.8	9.6	−4.9
S	25.9	17.4	13.6	−3.0
T	11.5	23.6	11.3	−2.0
B	15.0	20.0	18.0	−4.6
L	8.9	27.6	22.0	−4.3
M	9.1	14.9	10.1	−2.7
Q	3.6	3.3	24.3	−4.0

discriminate among products. Practical judgment therefore dictates that this factor is not meaningful.

Table 7 lists all of the factor scores as plotted in Figure 8. These were obtained using the highest loading variables as indicators of the factors. Table 7 makes another point clear about factor scores; they can be both positive and negative. A negative factor score does not mean that a product has a negative amount of an attribute. The negative values result from the fact that factor loadings can be negative. The sign only indicates which end of the variability continuum a particular product is representative of. For example, factor 1 ranges from a positive extreme of highly orange flavored to a negative extreme of more spicy than orange. All of the products in this study turned out to be more orange than spicy. However, if a product that was more spicy than orange were included, it would have a negative factor score on factor 1 because spicy is the negative extreme of this factor.

It is not necessary to conduct an entire new factor analysis to obtain factor scores of products not included in the original study. Such products need only be rated in the same fashion and on the same attributes as used in the original sensory study and factor scores computed from the obtained sensory ratings. Factor loadings obtained in the original factor analysis can be used as the weights for the sensory ratings. However, it is sometimes decided to use a weight of 1.0 for each attribute when products not included in the original study are described by factor scores. The rationale is that if these new products had been included in the original study, then the results of the analysis would have been different: while the same factors would likely have emerged, the loadings of each variable on the factors would likely be different because the additional products would have provided additional information to be accounted for by the correlations. Since there is no way of knowing precisely what those new loadings would be, a conservative approach is to weight all ratings the same at 1.0. In that case, the factor scores now become the simple sum of the attributes which are used as indicators of the factors. To facilitate comparison back to products contained in the original study, the weights applied to factor scores for those products ought to be changed to 1.0 as well.

The use of factor scores to summarize a data set does not preclude use of the original variables. First, the contribution of the original variables to the factor scores is reflected by the factor loadings, so each original variable is included to a known extent in the scores. Second, the original data is still available. Consequently, product manipulations can be carried out and documented using original sensory attributes as indicators, and the effect of the original attributes on the factor scores can be determined. The benefit of factor scores is that, being composite scores, they are more reliable indicators of a source of variability in the data than would be any one variable alone.

X. COMBINING CHEMICAL AND SENSORY MEASURES

Factor analysis and PCA can be used to determine relationships between sensory attributes and chemical components of the samples being studied. This is done by combining the sensory ratings and chemical measurements made on the samples into a single data set before performing any analyses. Since factor analytic procedures are based on correlations, it is not absolutely necessary to have the different types of data in the same metric. However, this can be accomplished if desired by transforming all data into z-scores. Each measurement, sensory as well as chemical, will then have a mean of zero and be in the same metric. The data can then be analyzed as discussed earlier. Two such analyses were performed on the data set being discussed here. The first was done on combined sensory and GC data collected on the tea samples, and the second was done on combined sensory and liquid chromatography (LC) data. The first analysis therefore examines the relationships among sensory ratings and the volatile components of the tea, and the second examines the relationships among sensory ratings and nonvolatile components.

Table 8 shows the results of a PCA performed on the combined sensory and GC data. The table contains the four components obtained after factor rotation. Examination of the first factor indicates that sensory ratings of orange flavors and aromas load highly on it as do chemical components likely to subtend such sensations. For example, ethyl butyrate, α-pinene, β-myrcene, ethyl hexanate, and octanal are all found in either orange oil or essence. The sensory ratings of clove and cinnamon aroma and taste all load highly on the second factor along with eugenol, a major flavor component of clove. The pattern of loadings suggests that the first factor is describing a component of variability made up largely of orange and related attributes, while the second factor describes a spicy component. The chemical compounds loading on these factors are those which contribute most heavily to these attributes.

The pattern of loadings on factors 1 and 2 are shown graphically in Figure 9, with the spice factor plotted against the orange factor. Since the chemical data are measurements of volatiles, the factors are labeled as aromas, but could just as well have been labeled as flavors. Spicy ratings and eugenol cluster together with high loadings on the spice factor and low loadings on the orange factor. Conversely, orange ratings and corresponding chemical compounds show the opposite pattern, i.e., high loadings on the orange factor and lower loadings on the spice factor. In this example it was known beforehand which compounds likely contributed to the spice and orange flavors. However, if this were not known, the clustering of chemicals and sensory ratings in the same area of the plot would indicate that the chemical measurements and the sensory ratings are correlated and would provide learning about the sensory-physical relationships in this category of beverage.

Table 9 and Figure 10 present the same type of analysis performed on the combined LC and sensory data. This analysis indicates relationships between soluble tea components and tea flavor. The first component of variability accounts for traditional tea flavors. Ratings of astringent and bitter are located in the same area of Figure 10 as are chemical components such as caffeine, gallic acid, epicatechin, epigallocatechin, and related compounds which taste primarily bitter and/or astringent. Ratings of "tea flavor" load in the same area as fumaric and succinic acid which are sour-bitter. While Figure 10 seems to indicate that "grassy flavor" loads very close to fumaric acid, this is actually not the case. Inspection of the loadings of "grassy" in Table 9 show that in fact it has a moderate negative loading on factor 3 and thus, in a three-dimensional space, would be situated fairly well behind fumaric acid. Orange and spicy flavors all tend to have relatively low loadings on factor 1. This is expected because the chemicals measured were solubles, and orange and spicy are largely due to volatiles. Factor 2 seems to account for sensory ratings of spicy flavors and factor 3 for sensory ratings of orange flavors. Sensory ratings that load highly on factor 2 or 3 tend not to load to a meaningful extent on factor 1 where solubles and the tastes/flavors they subtend cluster. An exception to this is sour taste which loads on factor 4 along with citric and malic acid. These two organic acids apparently account for much of the sour taste in the teas. The analysis therefore separated solubles responsible for bitter and astringent tastes (factor 1) from those responsible for a more pure sour taste (factor 4).

TABLE 8
Rotated Factor Loadings from Combined Gas Chromatography and Sensory Data

Variable	F_1	F_2	F_3	F_4	F_5	h^2
Ethyl butyrate	0.94	-0.30	-0.03	0.15	-0.02	0.99
α-Pinene	0.83	-0.13	0.43	0.08	-0.33	1.00
β-Myrcene	0.63	-0.09	0.72	0.10	-0.26	1.00
Ethyl hexanate	0.94	-0.30	-0.06	0.15	-0.03	1.00
Octanal	0.94	-0.12	0.26	0.10	-0.04	0.97
Limonene	0.35	0.01	0.92	0.04	-0.13	0.98
γ-Terpine	0.10	0.14	-0.14	0.24	-0.28	0.18
Linalool	0.92	0.01	0.26	0.18	0.09	0.95
Et-3-O-hexanate	-0.19	0.76	0.06	-0.07	-0.56	0.93
4-Terpineol	0.79	0.51	-0.23	0.16	0.07	0.96
α-Terpineol	0.47	-0.05	0.85	0.10	0.17	0.98
Neral	0.70	-0.17	0.68	0.10	0.01	0.99
D-Carvone	0.81	0.19	0.38	0.17	0.27	0.93
Geranial	0.78	-0.10	0.46	0.20	0.10	0.88
Cinnamic aldehyde	0.13	0.33	0.10	0.02	-0.88	0.91
Perialdehyde	0.34	0.35	0.63	0.18	0.20	0.70
Undecycaldehyde	0.24	0.19	0.31	0.36	0.32	0.42
Neryl acetate	0.08	0.04	0.97	0.03	0.19	0.98
Eugenol	0.02	0.96	0.20	0.02	0.12	0.97
Dodecanal	-0.04	0.08	0.99	-0.03	-0.04	0.99
β-Caryophyllene	0.14	0.13	0.98	-0.05	-0.09	1.00
Valencene	0.01	0.07	0.93	-0.07	-0.35	0.99
Orange aroma	0.64	-0.02	0.74	0.08	-0.13	0.98
Clove aroma	-0.19	0.94	-0.27	0.04	0.02	0.99
Cinnamon aroma	-0.25	0.88	0.00	-0.14	-0.36	0.98
Other aroma	-0.29	-0.64	-0.23	-0.33	0.32	0.75
Sweet taste	0.59	0.05	0.61	0.53	-0.08	1.00
Sour taste	0.52	-0.25	-0.06	-0.73	-0.33	0.97
Bitter taste	-0.25	0.25	-0.09	-0.80	0.00	0.77
Astringent taste	-0.05	0.15	-0.22	-0.87	-0.06	0.83
Tea taste	-0.50	-0.32	-0.14	-0.41	0.19	0.57
Candy-like orange	0.85	-0.18	0.46	0.17	-0.06	0.99
Fresh orange taste	0.53	0.32	0.60	0.33	0.16	0.87
Other fruit taste	0.27	-0.08	0.88	0.34	0.16	0.99
Clove taste	-0.11	0.98	-0.10	-0.04	-0.05	0.98
Cinnamon taste	0.04	0.94	0.18	-0.11	-0.04	0.93
Spicy taste	-0.39	0.14	-0.76	-0.40	-0.21	0.95
Grassy taste	-0.27	-0.41	-0.30	-0.44	0.18	0.55
Woody taste	-0.34	0.50	0.06	-0.74	0.10	0.92
Other taste	0.74	0.20	0.08	-0.30	-0.56	0.99
Overall preference	0.44	0.27	0.00	0.86	-0.01	1.00
Eigenvalue	11.26	7.23	10.78	4.96	2.67	
Eigenvalue/41	0.27	0.17	0.26	0.12	0.06	
Cumulative variance accounted for (%)	27	44	70	82	88	

The above two analyses combined sensory and chemical data by converting it to z-scores and then performing a single PCA on the two sets of data taken together as one. A logical extension of this approach is called partial least squares analysis.[8] In this approach, a PCA of a chemical-sensory data set is performed simultaneously with a regression analysis. The result is not only a set of factors, but a regression model in which one of the data sets is predicted from the other. For example, the analysis would provide a mathematical model which would allow a prediction of the sensory response that could be expected to result from a certain combination of the chemical compounds that were included in the analysis.

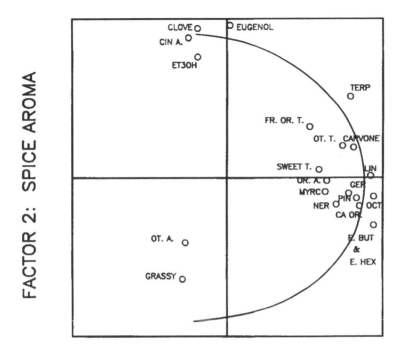

FACTOR 1: ORANGE AROMA

FIGURE 9.   Plot of rotated factors 1 and 2 derived from a principal components analysis performed on the combined sensory and GC measures made on the seven teas. Proximity of sensory and chemical variables in the plot suggest which compounds or groups of compounds are involved with determining specific sensory characteristics of the teas.

XI. ORTHOGONAL AND OBLIQUE ROTATIONS

The most common type of factor rotation is orthogonal. As discussed earlier, orthogonal rotations maintain the independence of the factors. When that is the case, the sum of the squared loadings of variables on each factor represents the amount of variability in the data set accounted for by the factor, and the sum of the squared loadings of any row in the factor table represents the amount of variability of that variable accounted for by the factors. This is a desirable situation because the degree to which the analysis describes the data can be ascertained.

In some cases an investigator will examine the geometric representation of factor loadings and decide that the axes of the plot could approach clusters of plotted variables more closely if the angle between the axes were something other than 90°. The axes are then rotated so that they pass through the clusters in such a way that the resulting angle between them is no longer 90°. This sort of rotation is referred to as an oblique rotation. The major benefit of oblique rotations is that loadings on factors are increased and clusters of variables may be more apparent. This is especially true in geometric representations where, with appropriate rotations, the axes of the plot can be made to pass directly through the middle of a cluster of variables. This is not always possible if orthogonal rotations are used.

When oblique rotations are used, the sums of the squared loadings in any column no longer represents the portion of variability accounted for by the factor. This is because oblique rotations result in factors which themselves are correlated. Consequently, some of the variability accounted for by one of the factors will also be accounted for by the others. The smaller the angle between the obliquely rotated factors, the larger the correlation will be between them

TABLE 9
Rotated Factor Loadings for Combined Liquid
Chromatography and Sensory Data

Variable	F_1	F_2	F_3	F_4	h^2
Gallic acid	0.97	−0.04	−0.18	0.00	0.97
Theobromine	0.97	−0.03	−0.14	0.19	0.99
Theophylline	0.08	−0.82	0.13	−0.42	0.87
EGC	0.89	−0.08	−0.31	−0.17	0.92
Caffeine	0.93	0.18	0.12	−0.27	0.98
Epicatechin	0.55	0.02	0.02	−0.68	0.76
EGCG	0.86	0.12	−0.13	−0.40	0.93
ECG	0.74	0.03	−0.24	−0.43	0.79
Citric acid	0.28	−0.31	0.32	−0.80	0.91
Malic acid	−0.67	−0.20	0.48	−0.50	0.96
Succinic acid	0.87	−0.45	−0.07	−0.08	0.97
Fumaric acid	0.69	−0.54	−0.20	−0.33	0.91
Orange aroma	−0.15	0.02	0.88	−0.37	0.93
Clove aroma	0.01	0.89	−0.26	0.13	0.87
Cinnamon aroma	0.35	0.90	−0.12	−0.07	0.95
Other aroma	0.48	−0.76	−0.38	0.19	0.98
Sweet taste	−0.26	0.01	0.93	−0.17	0.96
Sour taste	0.31	−0.25	−0.10	−0.84	0.87
Bitter taste	0.86	0.15	−0.39	−0.15	0.93
Astringent taste	0.75	0.05	−0.46	−0.37	0.91
Tea taste	0.76	−0.44	−0.40	0.25	0.99
Candy-like orange	−0.26	−0.23	0.78	−0.50	0.97
Fresh orange taste	−0.36	0.40	0.79	−0.27	0.98
Other fruit taste	−0.04	−0.09	0.97	0.17	0.97
Clove taste	0.06	0.96	−0.12	0.03	0.94
Cinnamon taste	0.03	0.96	0.14	−0.03	0.94
Spicy taste	0.31	0.13	−0.92	−0.13	0.97
Grassy taste	0.68	−0.57	−0.44	0.05	0.98
Woody taste	0.77	0.44	−0.28	0.03	0.86
Other taste	−0.04	0.24	0.24	−0.94	1.00
Overall preference	−0.54	0.17	0.49	0.00	0.56
Eigenvalue	10.96	6.50	6.65	4.53	
Eigenvalue/31	0.35	0.21	0.21	0.15	
Cumulative variance accounted for (%)	35	56	77	92	

and the greater will be the amount of common variability explained by both. Likewise, the sum of the squared loadings in a row no longer represents the amount of variability in the variable accounted for by the factors. In some cases of oblique rotation, the correlations between the factors themselves can be used as the basis for a factor analysis of the factors. This results in a set of higher-order factors that account for the relationships between the factors. Although there is nothing wrong with oblique rotations, and the results are often more pleasing in the sense that variable loadings are maximized on individual factors and minimized on all the others, the loss of information about variability accounted for must be kept in consideration.

XII. FACTORS VS. COMPONENTS

Throughout this chapter the terms "factor" and "component" have been used somewhat interchangeably. While this is a fairly common if somewhat loose and inaccurate practice, it is important to understand and keep in mind the distinction between these two terms. Theoretically, the term factor applies to the result of a common factor analysis. Recall that in common factor analysis the obtained factors explain only that variability which is shared in

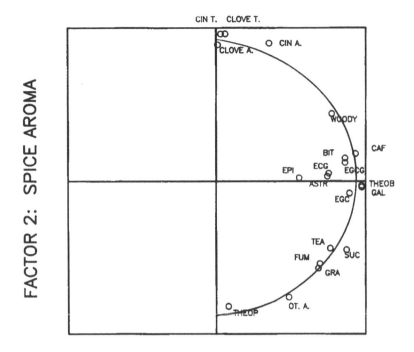

FACTOR 1: TRADITIONAL TEA FLAVOR

FIGURE 10. Plot of rotated factors 1 and 2 derived from a PCA performed on the combined sensory and liquid chromatography measures made on the seven teas.

common by sets of variables; the unique portion of the variability of each variable is not included in the analysis. In PCA both the common and unique portions of each variable are included in the results which are then referred to as components of variability of the data set. The main distinction between the two procedures is that a PCA is carried out on a correlation matrix which includes a value of 1.0 as each element of the main diagonal. This makes sense because the main diagonal represents correlations between each variable and itself and therefore 1.0 is proper. In common factor analysis, values other than 1.0 are inserted in the main diagonal. The values used are estimates of the communalities for the variables. This presents somewhat of a dilemma because there is no way of exactly determining communalities without first doing a factor analysis. Communalities are therefore usually estimated beforehand. For example, the squared multiple correlation between each variable and all other variables can be used as one estimate or the highest correlation that can be found between a given variable and all the other variables. However, these have a slightly different meaning than the communalities derived from a factor analysis which represent the variability in common between the variables and factors. Another approach is to use an estimate, such as the multiple correlation between a variable and all other variables, as an initial estimate of communality. An analysis is run using that estimate, and the communalities obtained from the factor analysis are then substituted in the original correlation matrix. A second analysis is then completed and the new communalities are inserted into the correlation matrix. The procedure is repeated until the communalities from successive analyses no longer change significantly. The final communalites will be less than one because only common variability is included in the analysis. PCA avoids this dilemma by including common and unique variability in the analysis. Consequently, communalities obtained in PCA are always 1.0. Whether one chooses to call the results components or factors is not as important as clearly understanding which portions of variability are being explained by each procedure.

REFERENCES

1. **Bennett, S. and Bowers, D.,** *An Introduction to Multivariate Techniques for Social and Behavioral Sciences,* Macmillan, New York, 1976.
2. **Kim, J. O. and Mueller, C. W.,** *Introduction to Factor Analysis: What It Is and How To Do It,* Sage, Newbury Park, 1978.
3. **Kim, J. O. and Mueller, C. W.,** *Factor Analysis: Statistical Methods and Practical Issues.* Sage, Beverly Hills, CA, 1978.
4. **Mulaik, S. A.,** *The Foundations of Factor Analysis,* McGraw-Hill, New York, 1972.
5. **Nunnally, J. C.,** *Psychometric Theory,* McGraw-Hill, New York, 1967.
6. **Rummel, R. J.,** *Applied Factor Analysis,* Northwestern University Press, Evanston, IL, 1970.
7. **Thurstone, L. L.,** *Multiple Factor Analysis,* University of Chicago Press, Chicago, 1947.
8. **Wold, S., Albano, C., Dunn, W. J., III, Esbensen, K., Hellberg, S., Johansson, E., and Sjostrom, M.,** Pattern recognition: finding and using regularities in multivariate data, in *Food Research and Data Analysis,* Martens, H. and Russwurm, H., Eds., Applied Science, New York, 1983.

INDEX

Printed and bound by CPI Group (UK) Ltd, Croydon, CR0 4YY

22/10/2024

01777633-0010